Modern Birkhäuser Classics

Many of the original research and survey monographs in pure and applied mathematics published by Birkhäuser in recent decades have been groundbreaking and have come to be regarded as foundational to the subject. Through the MBC Series, a select number of these modern classics, entirely uncorrected, are being re-released in paperback (and as eBooks) to ensure that these treasures remain accessible to new generations of students, scholars, and researchers.

T0184437

Perspectives on the History of Mathematical Logic

Thomas Drucker
Editor

Reprint of the 1991 Edition

Birkhäuser
Boston • Basel • Berlin

Thomas Drucker
Carlisle, PA 17013
U.S.A.

Originally published as a monograph

ISBN 978-0-8176-4768-1   ISBN 978-0-8176-4769-8 (eBook)
DOI: 10.1007/978-0-8176-4769-8

Library of Congress Control Number: 2007940495

Mathematics Subject Classification (2000): 01A05, 03-03

Cover design by Alex Gerasev.

Printed on acid-free paper.

9 8 7 6 5 4 3 2 1

www.birkhauser.com

Thomas Drucker
Editor

Perspectives on
the History of
Mathematical Logic

Birkhäuser
Boston Basel Berlin

Thomas Drucker
304 South Hanover Street
Carlisle, PA 17013
USA

Library of Congress Cataloging in Publication Data
Perspectives on the history of mathematical logic. Thomas Drucker,
    editor.
        p. cm.
    Includes bibliographical references.
        ISBN 0-8176-3444-4   ISBN 3-7643-3444-4
        1. Logic. Symbolic and mathematical—history. I. Drucker,
    Thomas.
        QA9.P43 1991   90-49462
        511.3'09—dc20   CIP

Typeset by Asco Trade Typesetting Ltd., Hong Kong.

9 8 7 6 5 4 3 2 1

ISBN 0-8176-3444-4
ISBN 3-7643-3444-4

Dedicated to
the memory of Jean van Heijenoort

Acknowledgments

I should like to thank all the authors for permission to include their contributions included here. Most of them were originally given as papers in the special session on the History of Logic at the American Mathematical Society's meeting in Chicago in March 1985. With regard to help in organizing that session, I am grateful to Ward Henson, Wolfgang Maass, and Anil Nerode. A number of papers that were delivered in that session, and are not included here, have appeared elsewhere or in some other form.

Thanks are due to John Dawson and to the Philosophy of Science Association for permission to reprint his paper, "The Reception of Gödel's Incompleteness Theorem," from the volume of proceedings of their annual meeting in 1984. It was felt that the paper's appearance here would help to secure for it an audience that might have missed its earlier publication.

I am grateful to the van Heijenoort family for permission to dedicate the volume to Jean van Heijenoort. The biographical tribute to him is by Irving Anellis.

Among the contributors I should like to express special thanks to Irving Anellis, William Aspray, and Judy Green. Anellis was responsible for the volume's having the chance to contain the three papers (Anellis on "The Löwenheim–Skolem Theorem," Seldin, and Wang) that were not delivered at the special session in Chicago, and for helping with the arrangements involved in dedicating the volume to his old teacher Jean van Heijenoort. Aspray supplied editorial experience, assistance, and judgment that were always welcome and helpful. Green furnished the impetus that was a *sine qua non* of the volume's appearance.

I should like to thank the staff of Birkhäuser Boston for their studious attention to the publication of the manuscript.

Contents

x   Contents

Contributors

IRVING H. ANELLIS
Modern Logic Publishing, Ames, Iowa 50011, USA

WILLIAM ASPRAY
Center for the History of Electrical Engineering, Rutgers University,
New Brunswick, New Jersey 08903, USA

JOHN W. DAWSON, JR.
Department of Mathematics, Pennsylvania State University, York,
Pennsylvania 17403, USA
and Institute for Advanced Study, Princeton, New Jersey 08540, USA

THOMAS DRUCKER
Department of Philosophy, University of Wisconsin, Madison,
Wisconsin 53706, USA

JUDY GREEN
Department of Mathematics, Marymount University, Arlington,
Virginia 22207-4299, USA

NATHAN HOUSER
Peirce Edition Project, Indiana University–Purdue University at
Indianapolis, Indianapolis, Indiana 46202, USA

STEPHEN C. KLEENE
Department of Mathematics, University of Wisconsin, Madison,
Wisconsin 53706, USA

DANIEL J. O'LEARY
37 Applevale Drive, Dover, New Hampshire 03820, USA

WIM RUITENBURG
Department of Mathematics, Statistics, and Computer Science,
Marquette University, Milwaukee, Wisconsin 53233, USA

Jonathan P. Seldin
Department of Mathematics, Concordia University, Montréal,
Québec H4B 1R6, Canada

Dirk Siefkes
Technische Universität Berlin, Institut für Software und Theoretische
Informatik, D-1000 Berlin 10, Germany

C. Smoryński
429 S. Warwick, Westmont, Illinois 60559, USA

Hao Wang
Department of Mathematics, Rockefeller University, New York,
New York 10021, USA

Jean van Heijenoort (1912–1986)

Irving H. Anellis

Jean van Heijenoort was formally trained as a mathematician, and specialized during the early part of his career in differential geometry. Most of his scholarly career, however, was devoted to logic, foundations of mathematics, and especially the history of logic, in which he was largely self-educated, but deeply influenced by his close friend Georg Kreisel.

From Frege to Gödel was his most influential work and the principal basis for his reputation. Yet it was not his only important contribution: he edited, as well, a collection of Herbrand's logical writings, served on a number of editorial boards, including that of the *Journal of Symbolic Logic*, to which he contributed many reviews, and, at the time of his death, was one of the co-editors of Gödel's collected works; there were, besides, his shorter papers in the history and philosophy of logic, some published during his lifetime, others not, many posthumously also published in his *Selected Essays*. In addition, he did original work, published and unpublished, in model-theoretic proof theory, centered on his proofs of the soundness and completeness of the falsifiability tree method for both classical and nonclassical logics.

For van Heijenoort (1976), "Mathematical logic is what logic, through twenty-five centuries and a few transformations, has become today." *From Frege to Gödel* documents the most crucial and profound of these transformations. Van Heijenoort's special interest within this perspective was the development of quantification theory as a family of formal systems providing different techniques for carrying out proofs in first-order logic. The principal contributors to this development were Frege and Hilbert, with their respective axiomatic systems and Hilbert's metamathematics; Löwenheim, whose work helped to clarify the concepts of satisfiability and validity for Hilbert's system; Herbrand, whose development of his own quantification theory unified and advanced the work of Hilbert and Löwenheim; Gentzen, who both developed the sequence calculus and helped formulate the method of natural deduction; and Gödel, whose incompleteness theorems defined the limitations of Hilbert's program. Van Heijenoort's booklet *El Desarrollo de la Teoria de la Cuantificación* encapsulates this history. Turning back from this perspective to the broad outline of the history of mathematical logic, van Heijenoort (1986) once wrote:

"Let me say simply, in conclusion, that *Begriffsschrift* (1879), Löwenheim's paper (1915), and Chapter 5 of Herbrand's thesis (1929) are the three cornerstones of modern logic."

References

van Heijenoort, J. (ed.) (1976), *From Frege to Gödel: A Source Book in Mathematical Logic*, 1879–1931, Harvard University Press, Cambridge, MA, Preface, p. vii.
van Heijenoort, J. (1986), Logic as Calculus and Logic as Language, *Selected Essays*, Bibliopolis, Naples (copyright 1985), pp. 11–16; see p. 16.

Introduction

Thomas Drucker

Mathematical logic is an impressive edifice, and the bulk of its development has taken place within the last century. Logic as a discipline goes back at least a couple of millennia further, at least as far as the formal representation of reasoning is concerned. Suggestions for a greater resemblance of logic to mathematics are frequent in the papers of Leibniz, and there is a sizable secondary literature which discusses what Leibniz may have had in mind. Nevertheless, the breathtaking rate of progress of mathematical logic, as attested to by the deluge of journals, texts, and papers, is a more recent development.

It is easy for a cursory survey of the development of the field to induce feelings akin to hero-worship. The 1930s were a period of creation in mathematical logic that continue to shape the field. Names such as Gödel (the subject of a couple of chapters in this volume) and Tarski, Church, Curry (the subject of another chapter herein), and Kleene (who is represented as author as well here) seem to have given mathematical logic an impetus it has never lost. We are fortunate to have some of these pioneers still with us, contributing technically, historically, and philosophically.

One of the remarkable contributions that these giants, upon whose shoulders later generations have stood, have made to our historical understanding is the extent to which their work arose out of discussions and in settings including colleagues less remembered today. The immense contributions of those named above and others emerged as the result of working on problems that were already in the air. The particular form their work took was in response to a particular teacher or exposition. There were shoulders upon which they too had stood in their turn.

The purpose of the history of science, broadly conceived, of which the history of mathematics and the history of mathematical logic form province and subprovince, is to understand the nature and development of the scientific enterprise. It seeks to replace the naive wonder at monuments with a sense for the tools and materials required for their creation. As the chapters here indicate, wonder does not die in the presence of understanding but becomes illuminated and illuminating. The practicing logician's ability to make progress thrives on an understanding of his predecessors. (I remember recently sitting at a dinner surrounded by logicians planning to spend a summer reading Schröder's 1890 lectures on the algebra of logic.)

The chapters included in this volume range broadly in time and subject, but they are all dedicated to unraveling the thoughts and the circumstances that have contributed to the evolution of mathematical logic. They involve technical details and philosophical underpinnings, support of colleagues and establishment of chairs. Some of the chapters give an insider's view of a particular development in the field, while others are detailed critical analyses of influential pieces of work. Their common feature is making sense rather than magic out of advances, binding together a community of contributors rather than leaving the impression of isolated wonder-workers. As the motto of the international chess federation has it, *Gens una sumus*. (We are one people.)

Bertrand Russell said that pure mathematics was born in the work of George Boole, whose work applied some of the techniques of algebra to the analysis of reasoning. It is safer to say that mathematical logic owes its birth to Boole's work, which gives the impression that progress has been made in Leibniz's dream of reducing argumentation to calculation. The kinds of laws being introduced in the study of algebra could be used at least to represent, if not entirely to supplant, the Aristotelean and subsequent rules governing the syllogism. The Latin tags associated with remembering which forms of the syllogism were valid could be replaced by algebraic argument, although logicians with a classical training did not immediately jump onto the mathematical bandwagon.

The point is that, with the work of Boole, questions about logic could take the form of mathematical questions, and Judy Green's paper on the problem of elimination looks at some nineteenth-century work in this setting. As she notes, the subsequent development of the field of logic has tended to cloak the contributions of logicians like Venn and Ladd-Franklin, and this has been unjust in two respects. In the first place, the work done on the problem has been ignored by those looking for progress in areas more easily recognizable. In addition, the background against which the logicians whose work remains better known has been inadequately represented. Green gives an introduction which offers the reader several pathways into the literature and incentives for pursuing them.

One of the figures whose contributions to the algebra of logic tend to have been ignored as part of the treatment accorded an unfashionable discipline was the polymath Charles Sanders Peirce. In Peirce's case his choice of notation and vocabulary has not aided the student seeking to understand his contributions to the development of logic. Rather more work is required, even by English-speakers, to get a handle on Peirce's methods and fundamental ideas.

Fortunately, Peirce has been the subject of attentive scholarship, exemplified by Nathan Houser's piece on the law of distribution. In addition to such published sources as the work of Carolyn Eisele, the archives of the Peirce project at Indiana University–Purdue University at Indianapolis have afforded Houser the material for pursuing Peirce's work on the law of distribution over a number of years. It should perhaps be recognized explicitly that many of the questions about the history of logic in which one comes to be interested are incapable of being answered in the absence of archival material on which to draw. Some of the other papers in this volume would have been impossible to write without the collections of the papers of logicians such as Russell and Gödel.

This observation can give rise both to melancholy and to hope. On the one hand, to quote Christian Thiel, "I must mention as a serious obstacle for the historiography of ... logic the loss of many Nachlässe of logicians during and even after the Second World War." This remark is found in his paper entitled "Some Difficulties in the Historiography of Modern Logic" in the *Atti del Convegno Internazionale di Storia della Logica* published in 1983. Although he relates some lugubrious anecdotes about the disappearance of archives, there is underlying the despair the hope that somewhere copies exist. Access to libraries in Eastern Europe and the Soviet Union may offer troves of material, even if finding out what is there can be time-consuming.

Fortunately, the Peirce archives have not had to suffer the effects of war the way many European archives have, and they furnish a good deal of illumination about the early days of axiomatic logic. The interplay between Peirce and Schröder bears witness to the stature of at least one representative of American mathematics in the European arena. Peirce was writing at a time when the direction of subsequent research in mathematical logic was far from clear, and some of his own shifts had their effect on the field. Houser's chapter rescues Peirce's work from its own notation and carries on the chronicle of the algebraization of logic.

There is no chapter in this volume exclusively devoted to the work of Gottlob Frege, but it is safe to say that there is no shortage of material elsewhere portraying a range of Fregean contributions. The two volumes of Michael Dummett on Frege, Hans Sluga's contribution to the series "The Arguments of the Philosophers," and Baker and Hacker's "excavations" present three approaches to Frege along quite different lines. In addition, Frege has left his traces on a number of other contributions, especially those concerned with Russell.

Bertrand Russell's technical contributions to mathematical logic came toward the earlier part of his career, but it is hard to overestimate the importance of *Principia Mathematica* in setting the tone for subsequent discussion of mathematical logic. Few readers may have read it cover to cover, but Russell and Whitehead's attitudes shaped the presentation of the discipline. It is not a coincidence that it was the system presented there that was the target of Gödel's work at the beginning of the 1930s, well after Russell had stopped actively pursuing mathematical logic.

Perhaps the best known incident in the history of mathematical logic was Frege's having given up in despair on his axiomatization of mathematics after receiving a note from Russell pointing out the derivability of a paradox (now known as Russell's) within Frege's formal system. Since this is a point at which Russell and Frege scholarship converge, there is no shortage of discussion of exactly the point Russell was trying to make and exactly how Frege took it. Irving Anellis uses material from the Russell archives at McMaster University to try to map out Russell's progress toward the result he sent to Frege, and its relationship to Cantor's original foundation of set theory. The result is a plausible reconstruction, originally from 1985, not entirely in line with suggestions put forward by those using other materials.

Anellis' chapter is an example of the care required to make progress in the history of mathematical logic rather than speculation. The besetting fault of the history of mathematics is to assume that there is only one path to a given result, and that it was pursued steadily and promptly once the result was in sight. Trying to document such a pursuit verges on the impossible and depends heavily on dating of letters, drafts, and even doodlings. There is a pleasure in reconstructing the development of an idea without worrying about finding evidence for it. Unfortunately, by the time evidence has been sought for and discovered, the pleasure has a good chance of giving way to fatigue. The reader making his way in the footsteps of the investigator may still find it a challenge, but the path will be easier to follow thereafter.

As an indication of the influence of *Principia Mathematica*, Daniel O'Leary's chapter discusses some of the early work on automated theorem proving. Since automated theorem proving falls within the fashionable discipline of artificial intelligence, there is no shortage of discussions of the methods themselves elsewhere. What O'Leary's chapter illuminates is the extent to which Russell and Whitehead furnished the model by which theorem proving is to be assessed. The manner in which Russell and Whitehead carried out their derivations could serve as a guide and a target for those wishing to find ways of carrying out the process mechanically.

Just as one cannot make bricks without straw, so one cannot do mathematics without money. Although it is not always clear how closely connected quantity and quality are in mathematical publications, there is no doubt that receiving attention from funding agencies will produce a flurry of activity in any mathematical discipline. Part, at least, of the enormous outburst of papers in mathematics after the Second World War can be put down to increased financial support.

William Aspray's chapter is a model of investigating the creation of a program in mathematical logic, namely, that at Princeton. He points out the wealth of accomplishment of those who spent time at Princeton in the 1930s and proceeds to explain how such a center could be created in so short a time. His work combines examination of university records and published sources with oral histories that he conducted to fill in the gaps in the available material. The result is a study of the interplay of campus politics and shifting intellectual alliances that leaves one amazed that constellations of genius are ever allowed to form by circumstance.

The historical dimension supplied by Aspray's chapter is one that historians of logic cannot ignore. Trying to explain the career of a logician of 50 or 100 years ago depends on the state of the discipline, but it also depends on the kind of welcome that the discipline received institutionally and financially. Getting a handle on the structure of decision-making within German universities before the First World War may not seem to have much to do with the development of mathematics, but it can be essential for understanding the careers of mathematical practitioners. Their work depended upon their surroundings, and their surroundings depended upon German educational policies and decision-making. Princeton in the earlier part of this century may not seem quite as far removed from the contemporary reader's experience, but the historian has the task of making sure that distinguishing features are not overlooked.

David Hilbert bestrode the German mathematical world like a colossus and when he developed an interest in mathematical logic, it received a good deal of attention. "Hilbert's Program" has been the subject of a good deal of literature, especially from the philosophical point of view, but the emergence of Hilbert's philosophical attitudes out of his mathematical practice is still in need of further attention. In general, the subject of proof theory was initiated not exclusively on a technical basis, but in order to see how far Hilbert's Program could be carried out. The results obtained did not always have a clear philosophical significance but initiated avenues for technical work.

Irving Anellis' second chapter ties together a number of the issues that attended the early years of proof theory. There has always been a kind of tension in the discipline between the figures of Herbrand and Gentzen, rendered romantic by their early deaths, and the details involved in assessing the consequences of adopting rules of inference in various forms. The discussion of proof theory ranges over a good part of the continent of Europe, England, and North America, all with the central goal of casting light on Herbrand's motivation for initiating such a fertile branch of mathematical logic.

In the same way that some notice was given to the absence in this volume of a chapter specifically on Frege, perhaps some excuse is appropriate for the absence of anything specifically devoted to the Polish school of logicians between the two wars. Kuratowski's volume of recollections has been translated into English, and even though it is devoted to Polish mathematics as a whole, logic bulked sufficiently large in the distinctive achievements of the community there to receive a good deal of attention. It is also worth mentioning the collection *Studies in the History of Mathematical Logic*, edited by Stanislaw J. Surma and published in Wrocław in 1973. Although not organized along national lines, the papers in that volume take up a number of themes and contributions of Polish logicians.

With the period 1930–1931 one arrives at the enigmatic and brilliant figure of Kurt Gödel, perhaps the most familiar of all mathematical logicians to the general population of mathematicians and even to the public. It seems that the complacency of the logical community after the appearance of

Principia Mathematica has been exaggerated, but the effects of Gödel's "Über formal unentscheidbare Sätze der *Principia Mathematica* und verwandter Systeme" have played their part across a variety of disciplines. Douglas Hofstadter's *Gödel, Escher, Bach* served in the 1980s to bring the name at least before the book-buying public (how many of the copies thoughtfully given during that decade were read with care is a question worth asking).

Oxford University Press has been publishing the collected papers of Gödel, and a couple of members of the editorial board of that project are contributors to this volume. John Dawson, in particular, has come to know the papers of Gödel (down to the handwriting) in a way that no one else does, and his chapter here draws attention to the role of Gödel's incompleteness theorems in the debates over the foundations of mathematics then prevailing. There has long been a tendency to come to Gödel's results in discussing philosophy of mathematics with a remark like, "What Gödel's theorems show is that Hilbert's Program was dead." Dawson points out that Gödel's technical results open doors rather than close them, and warns (as have some other recent commentators) about the dangers of attaching single philosophical labels to ideas from logic.

Two other chapters in this volume follow up the themes of Gödel and self-reference. Hao Wang spent a good deal of time talking with Gödel in his later years, and has transmuted some of the material of their conversations into a book of memoirs of Gödel. The wealth of Gödel's interests and the range of his speculations are sometimes easier to find in his conversations than to document in his published writings.

Wang uses his chapter here to speculate on the connection between ideas taking various forms. Even where problems are logically equivalent, that does not indicate that they can be solved with equal ease or that the consequences of their solution will be equally transparent. Logic traditionally has been connected with psychology, but logical equivalence cannot be replaced by psychological equivalence. The paper offers in practical form evidence of the importance of psychological contexts even in mathematical discovery.

The other chapter following from Gödel's work on self-reference follows the theme of self-reference rather than that of Gödel and comes from the pen of Craig Smoryński, who has already contributed to the literature of self-reference in a number of forms. There is a perennial fascination with the subject, since it can be captured with such ease with a single sentence and yet can produce such a divergence of proposed "solutions." Smoryński has long been concerned with keeping unjustified speculation separate from the evolution of mathematical logic, and his care in approaching the work of Hilbert, for example, has helped to keep the historical and philosophical record straight. Again the ease of coming up with tempting solutions to problems of self-reference makes it important to see how the consequences of self-reference have unfolded.

Smoryński's chapter shares with Wang's an interest in the different rates at which apparently equivalent fields and questions develop. He speculates on the sources of the differences when it comes to an area with such immediate philosophical attractiveness as self-reference. In the course of his paper he notes that writing history is neither sorting out credit nor preparing a chronology, but explaining a development. His use of technical developments in the field itself to explain the rate of development is ingenious, even if the ingenuity is a beneficiary of self-reference itself.

It would have been reasonable to say in the 1930s that a spectre was haunting European mathematics—the spectre of intuitionism. Although not many mathematicians may have felt tempted to follow the philosophical path that led Brouwer to intuitionism, they could appreciate the consequences for the use of certain results like the Axiom of Choice. The fear that intuitionistic methods would somehow deprive mathematics of results both valuable and beautiful was certainly a factor in Brouwer's being removed from the editorial board of *Mathematische Annalen* at Hilbert's instigation. The threat was seen as a political one as much as an intellectual one.

The danger to classical mathematics appears to have been greatly exaggerated, based on the quantity of mathematics being done and taught at present in an entirely classical style. Intuitionism has not, however, fallen into neglect. One source of the continued interest in Brouwer's ideas has been the attractions of constructive mathematics, whether from within mathematics or from computer science. Even those with little sympathy for Brouwer's philosophical position have to admit that nonconstructive methods are of little help to the programmer. Although the line of usefulness to the programmer would not be drawn quite where Brouwer had drawn it, the kind of ideas that Brouwer introduced would indicate the path to be followed.

More surprising perhaps is the appearance of intuitionism where one would not have expected it *a priori*. The whole subject of topos theory has provided a kind of algebraic support for intuitionism as a foundation for mathematics. This is but one of the lines of investigation examined by Wim Ruitenburg, who has carried on the tradition of Dutch scholarship in the area of intuitionism from the works of Brouwer, Heyting, and Troelstra. The idea of his chapter is to trace interpretations of intuitionism in unexpected places, and the widespread presence of these interpretations argues for something correct in the intuitionist attitude. Again technical details mingled with history provide better ammunition for a philosophical point of view than the speculation characteristic of earlier discussions of intuitionism.

Among the contributors to the articulation of the technical consequences of intuitionism, and the senior contributor to this volume, is Stephen Kleene, part of the glorious band at Princeton in the 1930s. Kleene's influence on the field of mathematical logic has been immense, both in terms of the books and papers he has written, and in terms of the students who have come out of the program at the University of Wisconsin. The Kleene Symposium held there in 1978 bore witness to the fruitfulness of his ideas in his own work and those of others.

For many years Kleene's *Introduction to Metamathematics* was the most widespread textbook for those interested in learning how logic worked. With revisions it continues to be popular in English and in translation, and here Kleene offers the story of how he decided what belonged in the book and in what order. Usually the analysis of a field in terms of its textbooks requires immense hours of reading and comparison, but the fact that Kleene took the trouble to put his own composition process on paper gives the student of the history of recent logic a head start.

Another of the great logicians of the 1930s, who remained active and influential throughout his life, was Haskell B. Curry. Curry's contributions to logic spanned a variety of areas, although his name is perhaps most associated with the ideas of formalism as a mathematical philosophy that he espoused in print. From the technical point of view, the area of logic for which he deserves a good deal of credit is that of combinatory logic, from which (in addition to the work of Church) many of the subsequent investigations of the lambda calculus take their start.

One memorial to Curry is the book *To Mock a Mockingbird* by Raymond Smullyan that came out in 1985 and was dedicated "to the memory of Haskell Curry—an early pioneer in combinatory logic and an avid bird-watcher." On a more modest scale is the chapter by Jonathan Seldin included here assessing the importance of Curry's work both for logic and for computer science. Seldin was both student and colleague of Curry and can serve as a guide through the thickets of combinatory logic.

A characteristic which has not worked to the disservice of mathematical logic in recent years has been its usefulness to the burgeoning discipline of computer science. It is often hard to tell whether the work done by those with a background in logic could not just as easily be called theoretical computer science at least. As noted above, intuitionism has been able to draw on the attractions of constructive mathematics for the computer scientist.

It is perhaps fitting then to end with a tribute to a logician whose work is of equal interest to workers in both logic and computer science. J. Richard Büchi worked in a variety of areas, and the kinds of conclusions to which he came (solutions to problems or measurement of levels of complexity) were diverse as well. Dirk Siefkes' chapter indicates the extent to which mathematical logic continues to offer techniques to other disciplines, as well as make progress on the questions associated with its past.

These chapters put together do not give a history of mathematical logic over the last century. Even on individual logicians they do not claim to be exhaustive. The collection does, however, indicate some of the richness of ideas that are involved in studying the history of logic and the variety of materials which can be pressed into service. National and chronological boundaries are not respected by the development of the field.

There are many areas in which progress can be made. Collected works for other logicians will offer the chance for commentary as well as for consultation. The interplay of political and social factors in the development of logic in Hungary, Czechoslovakia, and the Soviet Union, for instance, would make a fascinating narrative. The path back to Boole may not be as long as that in other branches of mathematics, but mapping out the complexities of the terrain will be a project of many years, tongues, and hands.

The Problem of Elimination in the Algebra of Logic

Judy Green

A central objective in any system of logic is to determine what conclusions follow from given premises. A special case is the recognition of valid syllogisms. Intermediate in generality is the problem of elimination, i.e., eliminating a logical variable from an equation or set of equations. The use of the word elimination in this context appears as early as 1854 in George Boole's treatise, *An Investigation of the Laws of Thought*, which contains the following description of the problem:

> As the conclusion must express a relation among the whole or among a part of the elements involved in the premises, it is requisite that we should possess the means of eliminating those elements which we desire not to appear in the conclusion, and of determining the whole amount of relation implied by the premises among the elements which we wish to retain. Those elements which do not present themselves in the conclusion are, in the language of the common Logic, called middle terms; and the species of elimination exemplified in treatises on Logic consists in deducing from two propositions, containing a common element or middle term, a conclusion connecting the two remaining terms. But the problem of elimination, as contemplated in this work, possesses a much wider scope. It proposes not merely the elimination of one middle term from two propositions, but the elimination generally of middle terms from propositions, without regard to the number of either of them, or to the nature of their connexion. To this object neither the processes of Logic nor those of Algebra, in their actual state, present any strict parallel.[1]

The elimination in "common Logic" to which Boole referred is exemplified by syllogistic reasoning, and the problem of elimination can therefore be interpreted as the search for an algorithm to extend reasoning of the syllogistic type to reasoning that involves more than two premises or three terms. Indeed, a derivation of some of the traditional syllogisms was quite generally included in the exposition of new algebras of logic.

Boole, in *Laws of Thought*, published a general solution to the problem of elimination of the term x from statements that can be written as equations of the form $f(x) = 0$, for functions $f(x) = ax + b\bar{x}$, where a and b are logical expressions in which x does not appear.[2] Boole correctly asserted and claimed to prove, but did not in fact prove, that any function of x can be written in the form $ax + b\bar{x}$.[3] Using that fact he showed that elimination of x from

$f(x) = 0$ produces the equation $f(1) \cdot f(0) = 0$.[4] In 1877 Ernst Schröder proved Boole's assertion about the representation of logical functions and modified Boole's formula for elimination to a more easily employable form, i.e., from $ax + b\bar{x} + c = 0$, which is equivalent to $(a + c)x + (b + c)\bar{x} = 0$, one deduces $ab + c = 0$.[5]

In 1881 John Venn characterized this solution of the problem of elimination as follows:

> We must say that the complete results of the elimination of any term from a given equation are obtained by breaking it up into a series of independent denials, and then selecting from amongst these all which either do not involve the term in question, or which *by grouping together can be made not to involve it* .... So understood, the rule for elimination in Logic seems complete.[6]

Despite Venn's statement of completeness for the rule of elimination and Boole's earlier claim "that in this peculiar system, the problem of elimination is resolvable under all circumstances alike,"[7] the Boole–Schröder method of elimination is not adequate even for a reasonable treatment of the syllogism. While the universal statements "All S is P" and "No S is P" have the respective algebraic translations $S\bar{P} = 0$ and $SP = 0$, the particular statements "Some S in P" and "Some S is not P" cannot be expressed as equations in Boole's system using only $S, P, \bar{S}, \bar{P}$, and 0. In order to represent particular statements, Boole introduced a symbol v that he described as "the representative of *some*, which, though it may include in its meaning *all*, does not include *none*."[8] With this interpretation of v, Boole expressed "Some S is P" as $vS = vP$ and "All S is P" as $S = vP$. He also reduced this representation of "Some S is P" to $vS\bar{P} + vP\bar{S} = 0$.[9] Although the introduction of the symbol v allowed Boole to derive the valid syllogisms, including those with a particular premise, he could not do so by simply applying his rule of elimination.[10] It was because of this difficulty of dealing with particular statements that a generally accepted solution of the elimination problem sufficient for a complete treatment of the syllogism came so late in the development of the algebra of logic.

In 1883, almost 30 years after Boole published *Laws of Thought*, Christine Ladd-Franklin introduced notation that led to a notable success in dealing with particulars.[11] Although Ladd-Franklin did not entirely succeed in solving the general elimination problem, she did solve the problem of the syllogism. She based her algebra on two copulæ: $\overline{\vee}$, representing mutual exclusion, and $\vee$, representing partial inclusion. Thus "No S is P" was represented by $S \overline{\vee} P$ or $(SP)\overline{\vee}$, while "Some S is P" was represented by $S \vee P$ or $(SP)\vee$.

This representation of the relations of traditional logic allowed Ladd-Franklin to give a simple solution of the problem of elimination for sets that consisted of premises of valid syllogisms. Since there are no valid syllogisms in which both premises are particular statements, her solution involved only two cases. Ladd-Franklin found that elimination of x from the two universal premises $a \overline{\vee} x$ and $b \overline{\vee} \bar{x}$ produced $(ab)\overline{\vee}$, while elimination from the universal premise $a \mathrel{\overline{\vee}} x$ and the particular premise $b \vee x$ produced $(b\bar{a}) \vee$.[12] The first of these is merely the Boole–Schröder method written using Ladd-Franklin's copulæ.

Using this restricted solution of the problem of elimination, Ladd-Franklin stated a recognition principle for all valid forms of the syllogism in terms of the two separate cases:

> All the rules for the validity of the doubly universal syllogism are contained in these:
>
> (1) The middle term must have unlike signs in the two premises.
> (2) The other terms have the same sign in the conclusion as in the premises.
>
> ...
>
> All the rules for the validity of the universal-particular syllogism are contained in these:
>
> (1) The middle term must have the same sign in both premises.
> (2) The other term of the universal premise only has its sign changed in the conclusion.[13]

Ladd-Franklin used the well-known fact that a syllogism is valid precisely if the set of three statements consisting of the premises and the denial of the conclusion cannot all hold at one time[14] to restate this recognition principle as follows:

> Take the contradictory of the conclusion, and see that universal propositions are expressed with a negative copula [$\overline{\vee}$] and particular propositions with an affirmative copula [$\vee$]. If two of the propositions are universal and the other particular, and if that term only which is common to the two universal propositions has unlike signs, then, and only then, the syllogism is valid.[15]

Her symbolic representation of this result, known as the Ladd-Franklin formula, is:

$$(a \mathrel{\overline{\vee}} b)(\bar{b} \mathrel{\overline{\vee}} c)(c \vee a)\overline{\vee}.^{16}$$

As noted above, the problem of elimination is considerably more general than just eliminating one of three terms from two statements, as in syllogistic reasoning. The general problem is to eliminate some among any number of terms from any number of statements, ending up with as much information as possible about all the remaining terms. In the same essay in which her treatment of the syllogism appears, Ladd-Franklin presented a solution to the problem of elimination for any number of conjunctions of universal statements and disjunctions of particular statements. Unfortunately, she did so using no symbols and left out an important clause of the conclusion. Furthermore, her solution was misinterpreted by Venn in his 1883 review of the book in which her article appeared.[17] Ladd-Franklin had not, in fact, stated a correct solution to the problem, but her omission was irrelevant to Venn's misinterpretation.

A universal statement, or conjunction (combination) of universal statements, can be written in Ladd-Franklin's notation as $(ax + b\bar{x} + c)\overline{\vee}$. A particular statement, or disjunction (alternation) of particular statements, can be written $(ex + f\overline{x} + g) \vee$. Ladd-Franklin wrote:

> From a combination of universal propositions, the conclusion, irrespective of any term or set of terms to be eliminated, x, consists of the universal exclusion of the product of the coefficient of x by that of the negative of x, added to the excluded combinations which are free from x as given.[18]

In symbols, eliminating x from $(ax + b\overline{x} + c)\overline{\vee}$ produces $(ab + c)\overline{\vee}$. This, as she pointed out, is just Schröder's modification of Boole's formula. Ladd-Franklin continued:

> If the premises include an alternation of particular propositions, the conclusion consists of the partial inclusion of the product of the total coefficient of x in the particular propositions by the negative of that of x in the universal propositions, added to the included combinations which are free from x as given.[19]

In symbols, eliminating x from $(ax + b\overline{x} + c)\overline{\vee}$ and $(ex + f\overline{x} + g) \vee$ produces the additional conclusion $(e\overline{a} + g) \vee$. This solution is incorrect in general since the additional conclusion should read $(e\overline{a} + f\overline{b} + g) \vee$.

Venn's mistake was based on his misinterpretation of Ladd-Franklin's use of the expression "included combinations which are free from x as given." Venn stated that Ladd-Franklin gave as the additional conclusion $(e\overline{a} + c + g) \vee$, indicating that he interpreted the term "included combinations" typographically to mean terms appearing in either the universal or particular statement.[20] In fact, Ladd-Franklin very clearly was using the term "included combinations" logically to refer to the terms appearing in the particular statement, or "inclusion," in the same way she had used the term "excluded combinations" to refer to the terms appearing in the universal statement, or "exclusion." Strangely, although Ladd-Franklin replied to Venn's review, she corrected neither his misinterpretation nor her own error.[21]

The misreading of Ladd-Franklin's solution by Venn is just one indication that he did not fully appreciate the importance of the concept of partial inclusion. While he did understand that Boole's use of the symbol v to represent particular statements was not completely successful, Venn was not at that time convinced, as Ladd-Franklin was, of the necessity of introducing a separate copula to represent such statements. He answered, "I am inclined to think that it is not" to the question of

> whether any perfectly general treatment of [particular propositions] is available, that is, corresponding in generality and brevity to those which Boole has given and which have been simplified in their practical employment by a succession of writers.[22]

It was not until logicians accepted a specific way of writing "Some S is P," usually as $SP \neq 0$ or $SP > 0$, that particular statements were treated systematically. For example, in 1894 Venn analyzed at length the symbolic expression $SP > 0$.[23] Although Venn's inclusion of this analysis in the second edition of *Symbolic Logic* indicates that he eventually understood the usefulness of a second connective to express particular statements, he continued to minimize the importance of such statements. In fact, his analysis was immediately followed by a discussion outlining the reasons for his belief that "particular propositions, in their common acceptance, are of a somewhat temporary and unscientific character."[24]

The difficulties connected with solving the problem of elimination for conjunctions of particular and universal statements, while based in part on a lack of reasonable notation for particular statements, were also rooted in the fundamental fact that a conjunction of particular statements is not equivalent to a single particular statement. Although Ladd-Franklin did not state a solution to the problem of elimination for conjunctions of both universal and particular statements, she did state:

> When a combination of particular propositions is included among the premises, the conclusion consists of a combination of the same number of particular propositions.[25]

She then gave as an example the elimination of x from $(px)\overline{\vee}, (qx)\overline{\vee}, (ax)\vee$, and $(bx)\vee$, which has as its strongest conclusion not involving $x, (a\overline{p}\,\overline{q})\vee$ and $(b\overline{p}\,\overline{q})\vee$. As $\overline{x}$ did not appear in any of Ladd-Franklin's premises, she neither repeated nor corrected the error she had made earlier in the essay.

In the same 1883 volume in which Ladd-Franklin's essay appeared, O.H. Mitchell introduced a new way of writing the universal statement "No S is P" as "Everything is either $\overline{S}$ or $\overline{P}$," $(\overline{S} + \overline{P})_1$.[26] Mitchell also introduced a new way of writing the particular statement "some S is P" as "Something is S and P," $(SP)_u$. He characterized his algebra of logic as one in which

> All propositions ... are expressed as logical polynomials, and the rule of inference from a set of premises is: *Take the logical product of the premises and erase the terms to be eliminated.*[27]

He noted that this rule works for all propositions, as long as no variable to be eliminated stands as a term alone. Mitchell pointed out that for universal statements this is equivalent to the negative (dual) of the Boole–Schröder method, i.e., the elimination of x from $(f(x))_1$ produces $(f(0) + f(1))_1$. Although Mitchell's method allows for the easy treatment of particulars, it makes the treatment of universals somewhat more complicated. For example, to handle the syllogism Baroko in Mitchell's system we translate the premises "All P is M" and "Some S is not M" as $(\overline{P} + M)_1 \cdot (S\overline{M})_u$. Since for any logical polynomials F and G, $F_1 G_u$ implies $(FG)_u$, the premises of Baroko imply $((\overline{P} + M)\cdot(S\overline{M}))_u$ or equivalently $(\overline{P}S\overline{M})_u$. Eliminating M by erasing, we get as our conclusion $(\overline{P}S)_u$, or "Some S is not P," as desired.

Mitchell also explained how his method could be used in other algebras of logic. Expressing "Nothing is F" by F_0, he noted that "F_0 and F_u are the two fundamental forms of proposition in [Ladd-Franklin's] method."[28] However, instead of explicitly solving the problem that Ladd-Franklin proposed, eliminating a variable simultaneously from a particular and from a universal statement, Mitchell merely stated that "*Elimination* from F_0 is performed by *multiplying* coefficients; from F_u, by *adding* them."[29]

The solution of the general elimination problem for conjunctions of universal and particular statements was first given by Schröder in 1885 when he showed that the elimination of x from $ax + b\bar{x} = 0, ex + f\bar{x} \neq 0, mx + n\bar{x} \neq 0,$ ... is $ab = 0, e\bar{a} + f\bar{b} \neq 0, m\bar{a} + n\bar{b} \neq 0, \dots .$[30] Both Ladd-Franklin's and Mitchell's systems were entirely adequate for the expression of the solution and either might easily have anticipated it. Indeed, as we have seen, Ladd-Franklin correctly presented an example that is equivalent to the elimination of x from $ax = 0, ex \neq 0,$ and $mx \neq 0$ but gave a flawed solution when there is only one equation and one inequality. In addition, she stated the solution in the special case when only the inequalities are present, i.e., when $a = b = 0$. Here again Ladd-Franklin did not use symbols to state the result, but this time her solution was correct, i.e., "From particular propositions by themselves no conclusion follows, otherwise than by simply dropping unnecessary information."[31] This is simply Mitchell's law of erasure applied individually to each particular proposition. Schröder acknowledged that this special case is implicit in Mitchell's work but did not mention that it is explicit in Ladd-Franklin's essay.[32] Had Mitchell, however, simply dualized $ax + b\bar{x} = 0$ to $((\bar{a} + \bar{x}) \cdot (\bar{b} + x))_1$, which equals $(\bar{a}\bar{b} + \bar{a}x + \bar{b}\bar{x})_1$, and applied his laws of multiplication and erasure, he would have anticipated Schröder's complete result. Like Mitchell himself, Schröder appears not to have seen the full implications of Mitchell's ideas.

Even after the problem of elimination was solved, it remained of interest to logicians. Schröder, in his lengthy *Vorlesungen über die Algebra der Logik*,[33] devoted many sections to the problem, and in 1901 Eugen Müller, citing Schroder's *Vorlesungen* as his source, published separately a detailed account of elimination and its application to the syllogism.[34] In a series of articles concerning logical equations, Platon Poretsky also addressed the problem of elimination.[35] In the first of these articles Poretsky derived both the Boole–Schröder law of elimination for $ax + b\bar{x} + c = 0$ and Mitchell's law of erasure for $ax + b\bar{x} + c = 1$.[36] He did not address the problem for particular statements $ax + b\bar{x} + c \neq 0$. Later he dualized the elimination problem in a novel way: he posed and solved the problem of finding the weakest statements not involving x from which each of $ax + b\bar{x} + c = 0$ and $ax + b\bar{x} + c = 1$ may be inferred. Poretsky called these statements *causes* of the original statements and asserted correctly that the cause of $ax + b\bar{x} + c = 0$ is $a + b + c = 0$, and the cause of $ax + b\bar{x} + c = 1$ is $ab + c = 1$.[37] However, Poretsky's proof relies on the inference of $a = 0$ or $b = 0$ from $ab = 0$. Unaccountably, Poretsky insisted on the universal validity of this inference even while quoting Schröder to the contrary.[38] It should be noted that N.I. Styazhkin in his *History of Mathematical Logic from Leibniz to Peano* claimed that "in the article 'The Law of Roots in Logic,'[39] Poretskiy solved the elimination problem for logical equations."[40] In fact, in that article, published in 1896 in Russian and in French, Poretsky solved the problem of finding expressions for one variable in a logical equation in terms of the other variables. Although in linear algebra solving for a variable is equivalent to eliminating it, this is not the case in the algebra of logic.

In his 1890 dissertation, A.H. Voigt also addressed the problem of solving a logical equation for an unknown.[41] In the course of doing so, Voigt factored $ax + b\bar{x} + c = 1$ as $(a + b + c)(x + b + c)(\bar{x} + a + c) = 1$ and obtained $a + b + c = 1$, reproducing Mitchell's law of erasure. He obtained a "general solution" of the original equation from the equations generated by the other two factors.[42]

Long after the algebra of logic had ceased to be an area of active research among mathematical logicians, David Hilbert and Wilhelm Ackermann returned to the subject, showing that it can be embedded in modern mathematical logic as the monadic predicate calculus.[43] Although they systematically derived the valid syllogisms within that context, their treatment masks the accomplishments of those logicians who developed the algebra of logic during the second half of the nineteenth century.

Notes

1. George Boole, *An Investigation of the Laws of Thought* (1854; reprint with corrections, New York: Dover Publications, 1958), 8.

2. Throughout this chapter we write $\bar{x}$ for $1 - x$ and adopt the convention that in any expression of the form $ax + b\bar{x} + c$, the only appearances of x are the ones explicitly shown.

3. Boole, *Laws of Thought*, 72.

4. Ibid., 101.

5. Ernst Schröder, *Der Operationskreis des Logikkalkuls* (Leipzig: B.G. Teubner, 1877: reprint, Stuttgart: B.G. Teubner, 1966), 14, 23.

6. John Venn, *Symbolic Logic* (London: Macmillan. 1881), 295.

7. Boole, *Laws of Thought*, 100.

8. Ibid., 124.

9. Ibid.

10. Ibid., 232–36.

11. Christine Ladd, "On the Algebra of Logic," in *Studies in Logic*, by Members of the Johns Hopkins University (Boston: Little, Brown & Co. 1883), 17–71. This essay was written before Ladd-Franklin was married and therefore it appeared under the name Christine Ladd. The essay was to have been Ladd-Franklin's doctoral dissertation at Johns Hopkins, but was not accepted as such because in 1882 Hopkins would not grant degrees to women. Ladd-Franklin received her Ph.D. from Hopkins in 1926.

12. Ibid., 38.

13. Ibid., 39. In 1883 G.B. Halsted pointed out a close relation between this disposition of the syllogism and an earlier paper of his own ("The Modern Logic," *Journal of Speculative Philosophy* 17 [1883]: 210–13; "Statement and Reduction of Syllogism," *Journal of Speculative Philosophy* 12 [1878]: 418–26). Ladd-Franklin's result was unquestionably independent and, in fact, was stated more clearly.

14. Ladd-Franklin later coined the term "antilogism" for this set of statements

("The Reduction to Absurdity of the Ordinary Treatment of the Syllogism," *Science* n.s. 13 [328] [12 April 1901]: 575).

15. Ladd, "Algebra of Logic," 41. E.W. Beth described Ladd-Franklin's work as "the first adequate treatment of classical syllogism" ("Hundred Years of Symbolic Logic," *Dialectica* 1 [1947]: 337).

16. Ladd, "Algebra of Logic," 40.

17. John Venn, review of *Studies in Logic*, by Members of the Johns Hopkins University, *Mind* 8 (1883): 594–603.

18. Ladd, "Algebra of Logic," 45.

19. Ibid., as corrected in the correspondence section of *Mind* 9 (1884): 322.

20. Venn, review of *Studies in Logic*, 599.

21. *Mind* 9 (1884): 322.

22. Venn, review of *Studies in Logic*, 598.

23. John Venn, *Symbolic Logic*, 2d ed. (London: Macmillan, 1894; reprint, Bronx, N.Y.: Chelsea Publishing Co., 1971), 184–88.

24. Ibid., 189.

25. Ladd, "Algebra of Logic," 46.

26. O.H. Mitchell, "On a New Algebra of Logic," in *Studies in Logic*, by Members of the Johns Hopkins University (Boston: Little, Brown & Co., 1883), 72–106. Mitchell's and Ladd-Franklin's representations of universal expressions are the only essentially new entries that Venn added to the list of translations of "No *S* is *P*" when he published the second edition of *Symbolic Logic* in 1894 (*Symbolic Logic* [1881], 407; *Symbolic Logic* [1894; reprint 1971], 481).

27. Mitchell, "New Algebra of Logic," 72.

28. Ibid., 97–98.

29. Ibid., 98.

30. Ernst Schröder, "Über das Eliminationsproblem im identischen Kalkul," *Tagblatt der 58. Versammlung deutscher Naturforscher und Ärtze in Strassburg* (1885): 353–54.

31. Ladd, "Algebra of Logic," 46.

32. Schröder, "Über das Eliminationsproblem," 353.

33. Ernst Schröder, *Vorlesungen über die Algebra der Logik*, 2d ed. 3 vols. (Bronx, N.Y.: Chelsea Publishing Co., 1966).

34. Eugen Müller, *Das Eliminationsproblem und die Syllogistik* (Leipzig: B.G. Teubner, 1901). Beilage zu dem Programm des grossherzoglichen Gymnasiums in Tauberbischofsheim für das Schuljahr 1900/1901. Programm Nr. 674.

35. Platon Poretsky, "Sept lois fondamentales de la théorie des égalités logiques," *Bulletin de la Société Physico-Mathématique de Kasan*, 2d ser., 8 (1898–99): 33–103, 129–81, 183–216; idem, "Quelques lois ultérieures de la théorie des égalités logiques," *Bulletin de la Société Physico-Mathématique de Kasan*, 2d ser., 10 (1900–01): 50–84, 132–80, 191–230; 11 (1902): 17–58.

36. Poretsky, "Sept lois fondamentales," 91–98.

37. Poretsky, "Quelques lois ultérieures," 10: 158–65.

38. Ibid., 10: 54.

39. Platon Poretsky, "La loi des racines en logique," *Revue de mathématiques* (*Rivista di matematica*) 6 (1896–99): 5–8; Russian version, "Zakon korney v logike," *Nauchnoe obozrenie* no. 19 (1896).

40. N.I. Styazhkin, *History of Mathematical Logic from Leibniz to Peano* (Cambridge: MIT Press, 1969), 219.

41. Andreas Heinrich Voigt, *Die Auflösung von Urtheilssystemen, das Eliminations-problem und die Kriterien des Widerspruchs in der Algebra der Logik* (Leipzig: Alexander Danz, 1890).

42. Ibid., 18–20.

43. D. Hilbert and W. Ackermann, *Grundzüge der theoretischen Logik* (Berlin: Julius Springer, 1928), 34–42; idem, *Principles of Mathematical Logic* (New York: Chelsea Publishing Co., 1950), 44–54 (translation of 2d ed. [1938], with revisions).

Peirce and the Law of Distribution

Nathan Houser

In the spring of 1880, not long before leaving for Europe on scientific business for the United States Coast and Geodetic Survey, Charles Peirce sent a paper entitled "On the Algebra of Logic" to *The American Journal of Mathematics*[1] (hereafter referred to as AJM–1880; I will similarly refer to two other papers published in that journal). Peirce had just taken up a lectureship at The Johns Hopkins University and in November 1879 wrote to his father that he was taking great pains with his lectures. But Peirce retained his assistantship in the Coast Survey and for the next 5 years conducted two careers, one as a lecturer in logic at The Johns Hopkins, the other as a scientist, already with 20 years of service for the Coast Survey, where since 1872 he had been in charge of measuring gravity. During those years Peirce was a frequent commuter on the B & O Railroad between Baltimore and Washington.

Peirce vigorously carried out his duties, but his health broke under the strain and during the winter and spring of 1880 he suffered several serious bouts of illness. That gave Peirce the opportunity to work on his logic, and in March, while confined to his quarters in Baltimore, he finished AJM–1880. In early April he submitted it for publication and by the end of the month was on his way to Europe on assignment for the Coast Survey, his mind primarily focused on scientific matters. The issue of *The American Journal of Mathematics* that carried Peirce's paper was printed in September 1880 and was in circulation by the end of the month. "On the Algebra of Logic" ran for over 40 pages and ended with a note that it would be continued. A paper of great complexity, it is described by Arthur N. Prior as an attempt by Peirce to bring syllogistic logic into "organic connection" with his logic of relatives.[2] It might be better described as part of Peirce's ongoing attempt to establish the adequacy of the algebra of logic as a general deductive logic, suitable for the representation of deductive thought in the broadest sense, by demonstrating its connection to syllogistic, the accepted standard of deductive inference for over two thousand years. In particular, Peirce hoped to show that the whole development of the algebra of logic could be based on syllogistic principles.[3]

Peirce began the formal presentation of his calculus with a section on "The

Algebra of the Copula" (Part 1, §4), in which he discussed the relation between logical consequence (premiss *therefore* conclusion) and implication (*if* antecedent *then* consequent) and obtained a reduction of syllogistic reasoning to Barbara.[4] He continued with "The Logic of Non-Relative Terms" (Part II, §1), his logic of "and" and "or."[5] This part of Peirce's logic, when combined with his algebra of the copula, gives a complete basis for and some development of classical propositional calculus.[6] It is not my purpose here to examine that achievement of Peirce, or to convey a sense of the full scope of his paper, which continues with his pioneering logic of relations.[7] My focus will be on just one of the theorems of Peirce's system, the law of distribution. My purpose is to give, in a summary (and preliminary) fashion, an account of Peirce's work on the law of distribution, and to say something about the results of his doubts about its status as a theorem. I will suggest that one result may have been Peirce's transition from his derivational approach to logic in 1880 to his truth-function approach of 1885.[8] In general, I hope to show that there was a major turning point in Peirce's logical development between 1880 and 1885 which calls for more concentrated attention.

In traditional logic "distribution" is likely to be understood with reference to the quantity of logical terms,[9] but here it is understood to concern logical operations, and the effect of applying operations to complex logical terms which themselves contain operational signs.[10] In particular, I am interested in the law of distribution (also known as the principle of distribution or the distributivity principle), which is the canonical expression of the relation of addition and multiplication or, in logic, disjunction and conjunction. In the algebra of mathematics the law of distribution is frequently given in the form

$$x(y + z) = xy + xz$$

and is called the principle of distribution of multiplication over addition. In symbolic logic, after an appropriate shift in the meaning of the connectives, that formula is called the principle of distribution of conjunction over disjunction. For logic Peirce added a second distribution principle,

$$(a + (b \times c)) = ((a + b) \times (a + c)),$$

which is called the principle of distribution of disjunction over conjunction.[11] Together, these two principles constitute what I call the full law of distribution. There are other distribution principles involving other operations, and different forms of the above formulas, but for my purposes there is no need for a wider account.

The mathematical principle of distribution must have been discovered by a pioneer of the virgin wilds of mathematics; it was enunciated in geometric terms by Euclid, and his proof is the first known proof of the principle.[12] According to Thomas L. Heath, in his commentary on Euclid, Heron of Alexandria was the first to adopt an algebraic form for the principle and to employ an algebraic method of proof. But Heron maintained that it is impossible to produce a strictly algebraic proof of the principle of distribution.[13]

Heron's claim appears to have been refuted in 1881 when the principle of distribution received an algebraic demonstration as a theorem of arithmetic at the hands of Charles S. Peirce. In "On the Logic of Number" (AJM–1881)[14] Peirce succeeded for the first time in constructing a simple set of axioms from which the natural numbers can be derived.[15] He proved the principle of distribution by mathematical induction, which he called Fermatian inference. However, the 1881 principle of distribution was strictly a principle of arithmetic and could not easily be generalized for logic. But Peirce had already made important advances in logic.

Concurrent with Peirce's interest in the foundations of mathematics was an equally deep and closely related interest in the foundations of logic. By the mid-1860s Peirce had become deeply involved in the movement to generalize algebra, a movement that had begun at least as early as Leibniz, and that blossomed in the work of George Boole.[16] Boole thought that the logic of Aristotle was not the foundational logic, that syllogistic operations were not the ultimate processes of logic.[17] He maintained that the ultimate laws of logic are mathematical in form and that logical method should be analogous to the method of mathematics. By the simple, yet ingenious, device of restricting the values of the terms of ordinary mathematical algebra to the two values 0 and 1, Boole discovered that numerical algebra can be transformed into a more general algebra of logic.[18] Boole recognized that this algebra of logic was subject to various interpretations and that it might be employed as a calculus of classes, of probability, or of propositions. He believed that the most distinguishing difference between his algebra of logic and ordinary algebra, resulting from the limitation to two values, was the validity of the index law $((a \times a) = a)$—$(a^2 = a)$—in the former but not in the latter.[19] Boole's disjunctive operation, symbolized by the sign of addition, was "exclusive," so that with reference to classes, *class* A plus *class* B equals *the class of everything* A *or* B *but not both* A *and* B. One important outcome of Boole's definition of "or" was the proscription of the following two formulas from his set of laws:

$$((a + a) = a),$$

$$((a + (b \times c)) = ((a + b) \times (a + c))).$$

The second of these formulas expresses the principle of distribution of disjunction over conjunction, which is invalid in Boole's logic as is its counterpart in arithmetic.

On 12 March 1867 Peirce presented a paper to the American Academy of Arts and Sciences[20] in which he introduced some modifications of Boole's calculus to improve its effectiveness for questions involving probability. Peirce clearly distinguished between arithmetical and logical operations, and employed distinctive signs to represent logical operators.[21] His most important deviation from Boole's system, a modification that Jevons had already made and that most followers of Boole have made,[22] was the substitution of inclusive for exclusive disjunction. That eliminated the barrier to a full duality between the laws pertaining to logical addition and multiplication.[23]

According to the principle of duality, for any theorem (or principle or law) in which the signs of conjunction or disjunction appear, there is a corresponding theorem in which the sign of disjunction replaces the sign of conjunction, and the sign of conjunction replaces the sign of disjunction, in every occurrence.[24] We saw that $((a \times a) = a)$ is a law in Boole's system, but its dual, $((a + a) = a)$, is not. Similarly, the principle of distribution of conjunction over disjunction, expressed by the formula $((a \times (b + c)) = ((a \times b) + (a \times c)))$, is valid in Boole's system, but its dual, the principle of distribution of disjunction over conjunction, expressed by the formula $((a + (b \times c)) = ((a + b) \times (a + c)))$, is invalid. But if we use Peirce's logical addition instead of Boole's exclusive disjunction, we can add $((a + a) = a)$ and $((a + (b \times c)) = ((a + b) \times (a + c)))$ to our logic and we are left with a thoroughgoing duality of laws.

Peirce's presentation on 12 March 1867 was the first introduction of the second part of the law of distribution. In his 1880 review of Frege's *Begriffsschrift*, Ernst Schröder remarked on the importance of that discovery and on Peirce's priority.[25] But Peirce not only asserted the full law of distribution for the Boolean algebra of classes, he also gave its first proof, which was published in 1868 in the Academy's *Proceedings*. I will not repeat the proof here—it can be found on pages 13–14 of the recently published second volume of the *Writings of Charles S. Peirce* and in *Collected Papers* 3.4—but will mention only that Peirce's method of proof was to show that the two main formulas of the law of distribution pick out the same region of the universe on each side of their equals signs. The proof is easily illustrated by Venn diagrams, a standard method of establishing the law of distribution in Boolean algebra today.[26] However, it is a method of proof limited to principles of the logic of classes.

In AJM–1880 Peirce again asserted the full law of distribution. His general procedure in presenting his 1880 calculus, a logic similar to systems of natural deduction,[27] was to prove his theorems from accepted argument forms and principles of inference. Peirce's intention was to restrict his basis to standard syllogistic forms and principles, with his definitions of × ("and") and + ("or") serving as key rules of inference. Unfortunately, Peirce did not explicitly set out the basis of his system.

Peirce's logic of nonrelative terms, starting with Part II, §1,[28] *appears*, at first anyway, to be self-contained. One notices immediately that there are numbered expressions (thirty-six in all) beginning with (1), and that most of the formulas after (3) are given as theorems. Peirce's expressions (1)–(3) are sometimes taken as the basis of his 1880 algebra of logic. They were presented by Peirce as follows (except for some notational changes),[29] where the symbol ∞ is to be regarded like the 1 of standard Boolean algebra, as representing the Universe (for Peirce the Universe of possibility), and 0 as representing the empty Universe (the impossible).

$$a \prec \infty, \qquad\qquad 0 \prec a. \qquad\qquad (1)$$

If $a \prec c$ and $b \prec c$,	If $c \prec a$ and $c \prec b$,	(2)
then $a + b \prec c$;	then $c \prec a \times b$;	
and conversely,	and conversely,	
if $a + b \prec c$,	if $c \prec a \times b$,	(3)
then $a \prec c$ and $b \prec c$.	then $c \prec a$ and $c \prec b$.	

From this basis, using uniform substitution, Peirce proved four theorems (actually theorem-sets) before asserting the law of distribution. His theorems may be given as follows:

$$a \prec (a + b), \qquad\qquad (4a)$$
$$(a \times b) \prec a, \qquad\qquad (4b)$$
$$b \prec (a + b), \qquad\qquad (4c)$$
$$(a \times b) \prec b; \qquad\qquad (4d)$$

$$a = (a + a), \qquad\qquad (5a)$$
$$(a \times a) = a; \qquad\qquad (5b)$$

$$(a + b) = (b + a), \qquad\qquad (6a)$$
$$(a \times b) = (b \times a); \qquad\qquad (6b)$$

$$((a + b) + c) = (a + (b + c)), \qquad\qquad (7a)$$
$$(a \times (b \times c)) = ((a \times b) \times c). \qquad\qquad (7b)$$

Peirce labeled (4a)–(4b) together as A; (5a) and (5b) together as B; (6a) and (6b), the *commutative principle*, together as C; and (7a) and (7b), the *associative principle*, together as D. He then asserted E, the *distributive principle* (the law of distribution),

$$((a + b) \times c) = ((a \times c) + (b \times c)), \qquad\qquad (8a)$$
$$((a \times b) + c) = ((a + b) \times (b + c)), \qquad\qquad (8b)$$

which he declined to prove, claiming it was easy but too tedious.[30] Although he had already published the 1867 distribution proof for the logic of classes, as a theorem of the logical calculus of AJM–1880, interpreted as a logic of propositions (and terms), the law of distribution stood in need of a rigorous new proof. Peirce clearly implied that his calculus was adequate for such a proof, and gave the distinct impression that he had produced one (or saw how to produce one). He apparently expected his readers to accept the derivability of the law of distribution on his word (or their acumen), but Peirce's assurance was not good enough for Schröder.

Sometime before 1880 Peirce and Schröder had become correspondents about the algebra of logic,[31] and it was not long after the publication of AJM–1880 that Schröder wrote to Peirce to ask about the omission of the proof.[32] Schröder conveyed his suspicion, if not already a conviction, that one of the principles of distribution was independent in Peirce's system. But Peirce was on assignment for the Coast Survey at the time of Schröder's inquiry and did not have access to his AJM–1880 notes; nor was he able to "reproduce" his proof. He concluded that he had been mistaken in thinking that he had produced a genuine proof, his mind having been clouded by illness when he wrote his 1880 paper.

In September 1883 Schröder presented a paper to the British Association for the Advancement of Science in which he claimed that Peirce could not have proved the law of distribution. (Schröder represented the law as four conditionals instead of two equivalences. When the principle of duality is applied to a conditional, in addition to the systematic replacement of the signs of conjunction and disjunction, the antecedent and consequent must be reversed.) Schröder contended that $((a \times (b + c)) \prec ((a \times b) + (a \times c)))$ and its dual, $(((a + b) \times (a + c)) \prec (a + (b \times c)))$, were independent in Peirce's algebra of logic and, therefore, could not be proved.[33] If those formulas were to be brought into logic, Schröder maintained, one of them would have to be taken as an independent axiom.[34] Thus $((a \times (b + c)) \prec ((a \times b) + (a \times c)))$ and its dual, $(((a + b) \times (a + c)) \prec (a + (b \times c)))$, were discredited as theorems of AJM–1880.

Schröder published his independence proof in 1890 in volume 1 of his *Algebra der Logik*. His proof, which was characterized by Huntington as succeeding "by a very complicated method,"[35] was widely accepted, initially even by Peirce.[36] But the independence of the law of distribution is *relative* to a system, in this case the system characterized by the basic expressions numbered (1)–(3). Given that basis, the full law of distribution cannot be derived, but, as Schröder maintained, can only be asserted as an independent principle.[37] However, a closer look at Peirce's logic of nonrelative terms suggests that (1)–(3) do not adequately represent its basis.[38] The most obvious deficiency is the absence of any formula involving negation. Although negative expressions are not asserted until (18), $(((c \times a) \prec b) \prec ((c \times \bar{b}) \prec \bar{a}))$, many follow, and it is clear that some basic negation formula is required. Of course, (18) comes too late to lend support to Peirce's distribution formulas.

To justify (18) Peirce appealed to the three clauses of his definition of negation from the section of AJM–1880 on the algebra of the copula (Part I, §4): (i) that $\bar{a}$ is of the form $(a \prec 0)$, (ii) $(a \prec \bar{\bar{a}})$, and (iii) $(\bar{\bar{a}} \prec a)$.[39] This crucial reference back to Part 1, §4 clearly suggests that Peirce intended the principles of his algebra of the copula to be applicable to his logic of non-relative terms. Furthermore, since all of the formulas of his nonrelative logic can be written as conditionals, it seems unlikely that he did not intend the basic principles of the algebra of the copula to carry forward.[40] While there are many principles or rules from Part I, §4 which *might* be appealed to in Part II, for present purposes I will select only the principle of identity

$$a \prec a \qquad\qquad \text{I, §4(1)}$$

and a principle that Prior describes as a relative of the deduction theorem, the principle that

$$\begin{array}{c} a \\ b \\ \hline \therefore\ c \end{array} \quad \text{and} \quad \begin{array}{c} a \\ \hline \therefore\ b \prec c \end{array} \qquad \text{I, §4(2)}$$

are of the same validity. As Prior points out, this latter principle resembles De Morgan's account of the role of implication in reasoning.[41] I will call this rule "Peirce's Rule," which I take to be strong enough to permit the inference of either side as conclusion from the other as premiss.[42] To elucidate my understanding of Peirce's Rule I will restate it in the following general form, where n may be any number greater than or equal to zero:

$$\begin{array}{c} A_n \\ B \\ \hline \therefore\ C \end{array} \quad \text{if and only if} \quad \begin{array}{c} A_n \\ \hline \therefore\ B \prec C. \end{array}$$

Peirce's practice suggests that something like this form is what he had in mind, although his statement of the rule appears to be the single case where $n = 1$. A complete account of Peirce's logic would certainly have to elaborate Peirce's justification for building up his algebra of the copula from this powerful principle (and the law of identity), but for the purposes of this paper I will admit the principle, I, §4(2), in its general form, and appeal to it to justify the assertion of rules as conditionals and the use of indirect proofs. This supposes that Peirce would have regarded the inference of $(a \prec b)$ from $(a \therefore b)$, or vice versa, as a case under the rule. Under the general form of Peirce's Rule, this would be the case where $n = 0$.[43]

Two other principles from Part I, §4, *modus ponens* and the principle of transitivity, will be employed, but they are not given as *basic* rules of inference. *Modus ponens* is obtained by substituting $(a \prec b)$ for a in (1), yielding $((a \prec b) \prec (a \prec b))$. Then by two applications of Peirce's Rule, we obtain (If $(a \prec b)$ and a, then b). The principle of transitivity is obtained by assuming $(a \prec b), (b \prec c)$, and a, from which b and c can be derived by *modus ponens*. By applying Peirce's Rule to discharge the third assumption, a, we obtain the transitivity principle, "If $(a \prec b)$ and $(b \prec c)$, then $(a \prec c)$." I will appeal to *modus ponens* (MP) and transitivity (Trans) in what follows as *derived* rules.

Instead of bringing in the negation formulas mentioned above (from Part I, §4), we can go to the beginning of Peirce's logic of nonrelative terms (Part II), where *before* he asserts (1) he gives the following as equivalent inferences (my numbers):

(i) $\begin{array}{c} a \text{ and } b \\ \hline \therefore\ c \end{array}$ is of the same validity with the inference $\begin{array}{c} a \\ \hline \therefore\ \text{Either } \bar{b} \text{ or } c \end{array}$

(ii) $\begin{array}{c} a \\ \hline \therefore\ \text{Either } b \text{ or } c \end{array}$ is of the same validity with the inference $\begin{array}{c} a \text{ and } \bar{b} \\ \hline \therefore\ c \end{array}$

He also equates the basic conditional $(a \prec b)$ with these two expressions:

(iii) (The possible) $\prec$ Either $\bar{a}$ or b;
(iv) a which is $\bar{b}$ $\prec$ (The impossible).

Selecting from these expressions I will augment Peirce's system by adding the law of identity (I, §4(1)), Peirce's Rule (I, §4(2)), and conditional forms of (ii) and (iii). The augmented system can be represented as follows:

R1. If $a, b \therefore c$, then $a \therefore (b \prec c)$, and the converse. (Peirce's Rule)
R2. If $(a \prec c)$ and $(b \prec c)$, then $((a + b) \prec c)$, and the converse.
R3. If $(c \prec a)$ and $(c \prec b)$, then $(c \prec (a \times b))$, and the converse.
  I. $a \prec a$.
 II. $(a \prec (b + c)) \prec ((a \times \bar{b}) \prec c)$.
III. $(a \prec b) \prec (\bar{a} + b)$.
IV. $a \prec \infty$.
 V. $0 \prec a$.

If we further augment Peirce's system to allow for indirect proofs, which supposes Peirce's Rule to be a form of the deduction theorem, we can prove the law of distribution (without using IV or V). To facilitate the proof I will use uniform substitution, as did Peirce, and rules associated with primary conditional formulas, and will permit the conjoining of properly obtained steps. These liberties can all be justified, to some extent at least, by Peirce's own procedure.[44] To facilitate the proof further I will make use of Peirce's theorems (4a)–(4d), *modus ponens* (MP), and transitivity (Trans).

1. $(a + b) \times c$.	Assumption.
2. c.	1, rule of (T4d) $\{S_{a+b}^{\ a}, \ _c^b\}$.
3. a.	Assumption.
4. $(a \times c)$.	3, 2 conjunction.
5. $(a \times c) + (b \times c)$.	4, rule of (T4a) $\{S_{a \times b}^{\ a}, \ _{b \times c}^b\}$.
6. $a \prec ((a \times c) + (b \times c))$.	3–5 discharge assumption (R1).
7. $\bar{a}$.	Assumption.
8. $((a + b) \times c) \prec (a + b)$.	$S_{a+b}^{\ a}, \ _c^b|$(T4b).
9. $(((a + b) \times c) \times \bar{a}) \prec b$.	8, rule of II $\{S_{(a+b) \times c}^{\ a}, \ _a^b, \ _b^c\}$.
10. $((a + b) \times c) \times \bar{a}$.	1, 7 conjunction.
11. b.	9, 10 MP.
12. $(b \times c)$	11, 2 conjunction.
13. $(a \times c) + (b \times c)$.	12, rule of (T4c) $\{S_{b \times c}^{\ b}, \ _{a \times c}^a\}$.
14. $\bar{a} \prec ((a \times c) + (b \times c))$.	7–13 discharge assumption (R1).
15. $(\bar{a} + a) \prec ((a \times c) + (b \times c))$.	14, 6 R2.
16. $(a \prec a)$.	I.
17. $(\bar{a} + a)$.	16, rule of III $\{S_a^b\}$.
18. $(a \times c) + (b \times c)$.	15, 17 MP.
19. $((a + b) \times c) \prec ((a \times c) + (b \times c))$.	1–18 discharge assumption (R1).

No effort has been extended to insure that the augmented system is minimal, and indeed if we admit the principle of duality, we see at once that it is not, for IV and V are duals and one can be dispensed with. Prior has shown further how to refine the set of fundamental principles of AJM−1880.[45] But the above set serves to show that the law of distribution is *not* independent in Peirce's 1880 logic relative to a carefully selected, but not unduly contrived, set of principles.

A different proof may be obtained without adding any negation formula if we strengthen R2 as follows (making it applicable to four terms):

R2:   If $((a \prec c)$ and $(b \prec c))$, then $((a + b) \prec c)$;
R2′: If $((a \prec c)$ and $(b \prec d))$, then $((a + b) \prec (c + d))$.

The proof runs as follows:

1. $a \times (b + c)$.		Assumption.
2. a.		1, rule of (T4b) $\{S_{b+c}^b\}$
3. $(b + c)$.		1, rule of (T4d) $\{S_{b+c}^b\}$
	4. b.	Assumption.
	5. $(a \times b)$.	2, 4 conjunction.
6. $b \prec (a \times b)$.		4–5 discharge assumption (R1).
	7. c.	Assumption.
	8. $(a \times c)$.	2, 7 conjunction.
9. $c \prec (a \times c)$.		7–8 discharge assumption (R1).
10. $(b + c) \prec ((a \times b) + (a \times c))$.		6, 9 R2′.
11. $(a \times b) + (a \times c)$.		3, 10 MP.
12. $(a \times (b + c)) \prec ((a \times b) + (a \times c))$.		1–11 discharge assumption (R1).

If we replace R2 with the complex constructive dilemma we can go directly from steps 3, 6, and 9 to step 11. A version of R2′ appears as theorem (12) in AJM−1880 and is easily derived in the augmented system:

1. $(a \prec c) \times (b \prec d)$.		Assumption.
2. $(a \prec c)$.		1, rule of (T4b) $\{_{((a \prec c),\ (b \prec d))}^a\}$.
3. $(b \prec d)$.		1, rule of (T4d) $\{_{((a \prec c),\ (b \prec d))}^b\}$.
4. $c \prec (c + d)$.		(4a) $S_c^a,\ _d^b$.
5. $a \prec (c + d)$.		2, 4 (Trans).
6. $d \prec (c + d)$.		(4c) $S_d^b,\ _c^a$.
7. $b \prec (c + d)$.		3, 6 (Trans).
8. $(a + b) \prec (c + d)$.		5, 7 R2.
9. $((a \prec c) \times (b \prec d))$ $\prec ((a + b) \prec (c + d))$.		1–8 discharge assumption (R1).

Peirce also asserted R2′ as a theorem in an early version of his paper.[46]

We see then that the independence of the principle of distribution depends on a restriction of the basis of Peirce's logic to I, R2, and R3; but I contend that such a restriction is by no means explicitly made by Peirce. It is a common complaint that Peirce often did not clearly lay out the foundations of his systems of logic.

The above proofs of the problematic part of the law of distribution are not Peirce's, and it is far from certain that he would have embraced any of them when he wrote his 1880 paper. It is, perhaps, less likely that he would have accepted them after Schröder's criticism which shook his confidence in the 1880 system and raised serious doubts that the algebra of logic could be developed in complete harmony with syllogistic.[47] For two thousand years syllogistic had been the accepted standard of deductive rationality. Even

Boole had *to some extent* judged the adequacy of his algebra of logic as a system of general logic by how it compared with syllogistic. It is not surprising that Peirce wanted to give his algebra of logic the pedigree of a syllogistic ancestry, and he thought he had succeeded in AJM–1880 until Schröder raised objections. Schröder convinced Peirce that part of the distributive law depends on principles that cannot be regarded as syllogistic and that any "proof" would have to appeal to *dilemmatic* principles of inference, which would annul Peirce's exclusive reliance on *syllogistic* principles.

The *dilemma* is a form of argument which appears to have roots in the disjunctive syllogism of the ancient Megarian and Stoic Schools that rivalled the Peripatetic School of Aristotle. It was first systematically employed as a form of inference around 1500.[48] Four standard dilemmatic forms were given by Richard Whately in his *Elements of Logic* (1826), from which Peirce got his first taste of Logic:

$$a \prec c$$
$$b \prec c$$
$$a + b$$
$$\therefore c$$

simple constructive dilemma

$$a \prec c$$
$$b \prec d$$
$$a + b$$
$$\therefore c + d$$

complex constructive dilemma

$$a \prec b$$
$$a \prec c$$
$$\bar{b} + \bar{c}$$
$$\therefore \bar{a}$$

simple destructive dilemma

$$a \prec c$$
$$b \prec d$$
$$\bar{c} + \bar{d}$$
$$\therefore \bar{a} + \bar{b}$$

complex destructive dilemma

Whately held that all reasoning whatever is fundamentally syllogistic and can be *reduced* to some set of standard syllogistic arguments, ultimately to Barbara. This is remarkably close to the view that Peirce held, and hoped to further, in AJM–1880.[49] According to Whately[50] a dilemma of this somewhat contrived form

$$p \prec q$$
$$\bar{p} \prec r$$
$$p + \bar{p}$$
$$\therefore q + r$$

can be reduced to these two conditional arguments

$$p \prec q \qquad \bar{p} \prec r$$
$$p \qquad \bar{p}$$
$$\therefore q \qquad \therefore r$$

Since we cannot deny both p and $\bar{p}$, we must admit one or the other of the conclusions, and thus we can simply substitute the two conditional syllogisms for the dilemma. But in Peirce's view there is a critical dependence here, and in any dilemma, on the law of excluded middle, which he came to regard as the characteristic feature of dilemmatic reasoning in general. In the definition of "dilemma" for the *Century Dictionary*, which Peirce contributed perhaps as early as 1886, he remarked that without the law of excluded middle the dilemma would not stand, but all of ordinary syllogistic would remain intact. Therefore the dilemma must be regarded as distinct from syllogism.

This is a rather thorny issue that needs to be thoroughly worked out. On the one hand, the law of excluded middle has long been regarded as one of the three traditional (and fundamental) laws of logic, the other two being the laws of identity and noncontradiction. On the other hand, Peirce came to regard the dilemma, at least partly *because* of its dependence on the law of excluded middle, as distinct from ordinary syllogism. Related difficulties with negation and reference to individuals crop up again and again in Peirce's writings.[51] But for now we must merely conclude that as a result of Schröder's criticism of AJM–1880, Peirce abandoned his attempt to "derive" his algebra of logic from syllogistic, having come to believe that an essential part of his logic, notably an important part of the law of distribution, depended on dilemmatic reasoning. At least by 1883 Peirce had definitely come to regard his 1880 calculus, in particular his definitions of logical multiplication and addition, as inadequate for the syllogistic foundation of the algebra of logic,[52] and by 1885 he openly credited Schröder with having revealed *this inadequacy*.[53] Yet Peirce never seemed reconciled to the expedient of simply asserting the problematic principle as an independent definition or axiom.

The period from 1880 to 1885 was the only extended period of Peirce's life when he was in regular dialogue with his own students of logic. There are nearly fifty writings on logic or mathematics in the Peirce Papers now assigned to those years, and it was in 1883 that *Studies in Logic*, by Peirce and his students, first appeared in print.[54] In 1883, when Peirce began to write "On the Algebra of Logic, Part II" (MS 527), which opens with a discussion of the faults of the first paper, he no longer seemed compelled to justify his basic principles by drawing them from a syllogistic foundation. Peirce experimented with alternative definitions of disjunction and conjunction in which logical values, or truth-values, appeared. It may have been a combination of Peirce's characterization of material implication ($a \prec b$) as true if either a is false or b is true, and false only if a is true while b is false,[55] and his full-fledged acceptance of dilemmatic reasoning, that suggested experimentation with truth-values; but in any case, Peirce was led to truth-function characterizations of his formulas, and suggested that $(a + b)$ be defined as $((a \prec b) \prec b)$ and, with "f" standing for "necessarily false," that $(a \times b)$ be defined as $((a \prec (b \prec f)) \prec f)$. By 1884, perhaps a year earlier, Peirce was using what is today commonly called the "method of assigning truth-values" as a decision procedure, and no longer thought that a logical truth required the pedigree of a syllogistic ancestry (or a derivation from laws of logic). At least by 1886 he was aware that the problematic part of the law of distribution could be shown, by the truth-value method, to be a logical truth.[56] It appears to have been this truth-function approach to logic that led Peirce, in 1883–84, to the discovery of the principle now known as Peirce's law, $(((a \prec b) \prec a) \prec a)$, the fifth icon in his well-known 1885 paper on the algebra of logic (AJM–1885).[57] In a footnote to AJM–1885 Peirce associated his "fifth icon" with dilemmatic reasoning, remarking that the dilemma "involves" that formula. His idea may have been that the necessity of his fifth icon can be demonstrated by arguing dilemmatically that whether a is true or false the icon will be true so it must be necessary. He explained that in AJM–1880 he had overlooked the distinctive character of the dilemma, and that that oversight, probably due to the general neglect of that mode of reasoning, had led him to the erroneous belief that "the whole of nonrelative logic was derivable from the principles of ancient syllogistic."

It is somewhat curious that after having recognized the special character of dilemmatic reasoning and the need for including dilemmatic principles with syllogistic principles for a complete calculus of logic, Peirce did not return to the logic of AJM–1880 to see what could be salvaged. His "failure" to produce an adequate proof of the law of distribution when questioned by Schröder, and his apparent acceptance of Schröder's claim that it was impossible in the 1880 system, seem to have resulted in his loss of confidence in that system. Consequently, it must have come as a shock when in turning over old papers from the 1880s he discovered his long lost proof, written up as though for publication. We do not know when Peirce recovered his proof but the story did not begin to emerge until the latter days of 1903.

On 11 December of that year Edward V. Huntington wrote to Peirce that he was "engaged in working out the independence of the postulates which lie at the basis of the algebra of symbolic logic" and that he wanted to discuss some "questions of notation" before writing up his paper in full for presentation to the American Mathematical Society.[58] A card from Huntington dated 29 December 1903 indicates that Peirce had replied with a painstaking letter and had taken the opportunity to reveal his recovered proof. Huntington wrote that he would "go over the proof with the greatest interest." (Peirce's letter to Huntington containing the proof is not extant, but in the Peirce Papers at Harvard there are several preliminary drafts, all dated 24 December 1903, of the proof and part of Peirce's letter.[59]) The crucial principle of that proof, not explicit in AJM–1880, is identified by Huntington as

If $a \prec b$ is false,
    then there is an element x, not 0, such that $x \prec a$ and $x \prec \bar{b}$.

After examining Peirce's proof, Huntington wrote back to proclaim its soundness and to acknowledge its importance.[60] He said that by using the proof he could "decidedly" simplify his list of assumptions, and he asked for Peirce's permission to use the demonstration. Peirce was pleased to comply, and when

Huntington's paper appeared it contained a lengthy footnote to Theorem 22a giving Peirce credit for the demonstration and reproducing a note by Peirce that gave a brief history of the proof.[61] Thus, nearly a quarter of a century after it should have appeared, Peirce's proof of the problematic part of the law of distribution came into the mainstream of the development of formal logic. When C.I. Lewis published his *Survey of Symbolic Logic* in 1918, he used Peirce's proof which he took from Huntington.[62]

Peirce's omission of the proof in 1880, and his subsequent inability to produce it for Schröder, may have raised some doubts that there ever had been a proof, even one that Peirce had mistakenly thought sound. But when Huntington published his landmark paper in 1904, such suspicions were soundly put to rest; clearly Peirce *had* a proof. The employment of Peirce's proof by Huntington and Lewis is testimony to its soundness and importance, and its validity has recently been endorsed by Roberts and Crapo in their paper entitled "Peirce Algebras and the Distributivity Scandal," presented in 1968 to the Association for Symbolic Logic.[63] Still it might be asked whether there is any evidence, beyond Peirce's own account, that he had the proof in 1880 when he published his paper, or whether he produced it later to relieve embarrassment.

In the Peirce Papers at Harvard there is a note to AJM–1885 reported by the editors of the *Collected Papers* to have been written up for publication shortly subsequent to that paper.[64] In the note Peirce outlined part of a proof of the problematic principle of distribution (though not the most problematic part) that is remarkably similar to part of the proof that he sent to Huntington. It might be conjectured that Peirce did not produce the "Huntington" proof before he wrote his 1885 paper, and that the proof he uncovered is the one he describes in his AJM–1885 note.

But even if the proof Peirce sent to Huntington was not originally drawn up for his 1880 paper, he was not mistaken in thinking he had produced a proof for that paper. A recent study of a little-examined manuscript, MS 575, has revealed that it is almost certainly part of an early version of AJM–1880. More importantly, it contains a somewhat abbreviated proof of the full law of distribution from the following principles:[65]

R1. If $(a \prec b)$ and $(b \prec c)$, then $(a \prec c)$.

R2. If $(a \prec c)$ and $(b \prec c)$, then $((a + b) \prec c)$.

R3. If $(c \prec a)$ and $(c \prec b)$, then $(c \prec (a \times b))$.

  I. $(a + b) \prec (b + a)$.

 II. $(b \times a) \prec (a \times b)$.

III. $a \prec (a + b)$.

IV. $(a \times b) \prec a$.

 V. $a \prec 1$.

VI. $0 \prec a$.

VII. $a \prec a$.

VIII. $b \prec ((a \times b) + (\bar{a} \times b))$.

 IX. $((a + b) \times (\bar{a} + b)) \prec b$.

There is a close resemblance between these expressions and those given in AJM–1880; only R1 and the negation formulas, VIII and IX, do not appear in the AJM–1880 list of basic principles or theorems (see page 14 above). R1 appears in AJM–1880 as a derived rule, but some strengthening of the basis as given on page 17 is required before VIII and IX can be derived. This might be achieved in various ways, for example by adding the principle of double negation, a form of which was given by Peirce in AJM–1880 (Part 1, §4) as part of his definition of negation, and by asserting III (page 17), following Peirce, as an equivalence instead of a conditional. Or we might simply replace R2 with R2′ (page 18), although R2′ was not given in AJM–1880.

My concern in this chapter has been to consider Peirce's systems only in connection with the law of distribution, not in connection with broader concerns such as completeness, consistency, and minimality. A comparison of MS 575 and AJM–1880 from this broader perspective would shed light on the development of Peirce's logic in the few weeks, or at most months, that separate them. A preliminary point of interest is that the MS 575 system does not include Peirce's Rule (R1 from AJM–1880) without which we can neither use an indirect method of proof, nor derive the rule of *Modus Ponens*.

I have reconstructed Peirce's MS 575 proof (without using V or VI) and can attest that it is *tedious*, as Peirce claimed it to be in AJM–1880, for my reconstruction runs to over 150 steps. It is not the same proof that Peirce sent to Huntington in 1904. Nevertheless, it is now clear that when Peirce sent his 1880 paper to *The American Journal of Mathematics* he had indeed produced a proof of the full law of distribution.

But even if Peirce's proof should turn out to be flawed, a new and careful look at Peirce's 1880 paper and the transitional writings between it and AJM–1885 is desirable, for during that 5-year period Peirce introduced two major discoveries into the history of logic: truth-function analysis and the logic of quantification.[66] It is likely that Peirce's development of quantification theory, stimulated by the conceptual and notational advances of his student O.H. Mitchell,[67] is in part the result of his effort to overcome Schröder's criticisms. To some extent, then, both of these important developments grew out of Peirce's dissatisfaction with his 1880 paper. Moreover, it was during those years that Peirce was finally able to shake off the firm grip of syllogism, a liberation that may be taken to represent his decisive turn to modern logic.

Notes

For helpful comments or advice I wish to thank James Van Evra, Max H. Fisch, Christian J.W. Kloesel, Ursula Niklas, Paul B. Shields, and Atwell Turquette. Special thanks go to Don D. Roberts for his astute and gracious assistance. The Department of Philosophy of Harvard University has generously granted me permission to refer to and quote from their collection of Peirce Papers, now housed in the Houghton Library.

A preliminary version of this chapter was presented in March 1985 to The American Mathematical Society in a Special Session on the History of Logic, organized by Thomas L. Drucker.

1. "On the Algebra of Logic," *American Journal of Mathematics* 3 (1880), 15–57. "On the Algebra of Logic" is included in the *Collected Papers of Charles Sanders Peirce* (CP 3.154–251) and will be published in volume 4 of the *Writings of Charles S. Peirce*.

Throughout these notes abbreviated references will be made to four sources: (1) *Collected Papers of Charles Sanders Peirce*, vol. 3, eds. Charles Hartshorne and Paul Weiss, Cambridge, Mass., Harvard University Press, 1933; (2) *Writings of Charles S. Peirce*, vol. 2, eds. Moore, Fisch, Kloesel, Roberts, and Ziegler, Bloomington, Indiana University Press, 1984; (3) *New Elements of Mathematics by Charles S. Peirce*, vol. III-1, ed. Carolyn Eisele, The Hague, Mouton, 1976; and (4) The Charles S. Peirce Papers at the Houghton Library of Harvard University. In references to the *Collected Papers* I will give volume and paragraph numbers (CP 3.xyz), for *Writings* I will give volume and page (W2.xyz), for *New Elements*, volume and page (NEIII-1.xyz), and for the Peirce Papers I will give manuscript numbers according to the *Robin Catalogue* and page numbers as assigned by the Institute for Studies in Pragmaticism, Lubbock, Texas (MS xyz, p. #).

2. Arthur N. Prior, "The Algebra of the Copula," *Studies in the Philosophy of Charles Sanders Peirce*, Second Series, eds. Edward C. Moore and Richard S. Robin, University of Massachusetts Press (1964), 79–94. In Prior's opinion Peirce did not quite achieve his purpose, although he did take a definite step forward.

3. Peirce's comparison of syllogistic with Boole's algebra of logic was underway by 1865 when he gave a series of lectures at Harvard entitled "On the Logic of Science." Those lectures appear in the first volume of the new edition of Peirce's writings, *The Writings of Charles S. Peirce 1857–1866* (see especially the sixth lecture, pp. 223–239).

4. This may overstate Peirce's achievement. Peirce is well known for having shown as early as 1866 in his "Memoranda Concerning the Aristotelean Syllogism" (W1.505; CP 2.792–806) that every reduction of one syllogistic figure to another takes the logical form of the figure being reduced. Max Fisch calls this Peirce's "first major discovery in logic" (W1.xxxv). However, Prior claims that Peirce's AJM–1880 reductions are free of implicit assumptions of what is to be proved. (Prior, 84.)

5. The original publication (AJM–1880) was divided into "Chapters" instead of "Parts."

6. See Prior for an account of the success of Peirce's 1880 logic. It is not universally agreed that Peirce's purpose was to construct a propositional logic *per se*. For criticism of Prior's account see Randall R. Dipert, "Peirce's Propositional Logic," *Review of Metaphysics* 34 (March 1981), 569–595.

7. Peirce used the term "relative" instead of "relation," but J. Brunning, in a recent dissertation for the University of Toronto has forcefully argued that Peirce knew he was working with *relations* in 1880 (even as early as 1870).

8. In 1880 Peirce thought that to establish theoremhood a statement had to be *derived* by accepted rules from pedigreed laws (in 1880 the best pedigree from Peirce's standpoint seems to have been an undisputed descent from syllogistic), while in 1885 he admitted as "theorems" (logical truths) statements shown to be *necessary* by truth-function analysis. Without explicitly saying so, Peirce seems to have expanded his conception of *validity* to include what is now often called "semantic validity." See Susan Haack, *Philosophy of Logics* (Cambridge University Press, 1978, pp. 13–14) for a brief account of the syntactic–semantic distinction.

9. After Peter of Spain a term is said to be distributed when it is taken universally and to be undistributed if it is not taken universally. In the statement "All men are

mortal," the term "men" is distributed because its reference is to every man, but "mortal" is undistributed because its reference is not to every mortal. In general, for traditional Aristotelean logic, we can classify the four propositional forms by the status of their subject and predicate terms with reference to distribution. (Baldwin's *Dictionary of Philosophy and Psychology*, Vol. 1, "Distribution," by R. Adamson, p. 289.)

10. Boole gives an early *general* definition of distribution on p. 34 of *The Laws of Thought* (Dover, 1958). See also *The Mathematical Analysis of Logic* (1847), pp. 16–17.

11. Usually I will use lowercase letters from the front of the alphabet (a, b, c, d) to represent logical terms, and lowercase letters from farther back (x, y, z) to represent numerical terms. Peirce's usage varies and my "standardization" of notation may put the reader at risk of missing certain subtleties of Peirce's work. However, in a paper such as this one, there is at least an equal risk of confusion likely to result from an erratic and inconsistent notation.

12. Euclid's principle (II, 1) was given as follows: "If there be two straight lines, and one of them be cut into any number of segments whatever, the rectangle contained by the two straight lines is equal to the rectangles contained by the uncut straight line and each of the segments." (*Euclid's Elements*, Dover, I, 375.)

Suppose we have a rectangle with one side of length x and with an adjacent side of length w. For the simplest application of Euclid's principle we shall suppose that side w is divided into two segments of lengths y and z. Thus $w = y + z$. According to Euclid's principle the area of the rectangle xw is equal to the sum of the areas of the rectangles xy and xz. Since $w = y + z$, we can characterize this simplest case of Euclid's principle as $x(y + z) = xy + xz$, which is the law of distribution for the algebra of mathematics.

Euclid's demonstration of II, 1 is given and discussed by Hogben in *Mathematics for the Million*, 135ff. In 1888 David Hilbert published his *Grundlagen* in which he, following others, completed Euclid's *Elements*. Hilbert's proof of the distributive law for multiplication appears in Beth's *The Foundations of Mathematics*, p. 146.

13. Heron lived in the third century A.D. and wrote a systematic commentary of the *Elements*. See Heath, Dover, 20ff., for an examination of the evidence for dating Heron's commentary. Heath indicates that Heron expressed the law of distribution in a formula equivalent to the standard algebraic expression $a(b + c + d) = ab + ac + ad$. Heath attributes this information to the commentary of an-Nairizi (ed. Curtze, p. 89). See Heath, Dover I, 373. However, the signs " $+$ " and " $\times$ " as symbols of addition and multiplication did not make their appearance until about the fifteenth century, and there do not seem to have been any previous commonly accepted symbols for these operations. See Hogben, *Mathematics for the Million*, pp. 306–307.

14. *American Journal of Mathematics* 4 (1881), 85–95; CP 3.252–288.

15. For the demonstration of Peirce's priority in this achievement see Paul Bartram Shields's dissertation, *Charles S. Peirce on the Logic of Number* (Fordham University, 1981).

16. There are scores of important mathematicians and logicians that could be located between Leibniz and Boole. Some giants stand out; Augustus De Morgan and Sir William Rowan Hamilton perhaps above the rest; but there are many others, not the least of which was Charles Peirce's father, Benjamin Peirce. Yet it is generally agreed that from the standpoint of logic George Boole played the greatest single role in the turn to modern thought.

17. Boole, *The Laws of Thought*, p. 10.

18. Boole, *The Laws of Thought*, p. 37; see also p. 166.

19. See Boole's *The Mathematical Analysis of Logic*, p. 18, and *The Laws of Thought*, p. 37. This special law of logic appears to have been recognized by Leibniz: see his "Study in the Logical Calculus," written in the early 1690s, and published in Loemker's *Philosophical Papers and Letters of Leibniz*, pp. 380–81.

20. Peirce's paper was published early in 1868 by the Academy though he had preprints by December 1867.

21. Peirce introduced the sign "$=$" to signify numerical *identity* (distinguished from numerical *equality*); thus "*a* logically equals *b*" ($a = b$) means "*a* and *b* denote the same class." He used a comma to signify logical multiplication (conjunction) and a plus sign with a comma in the lower right-hand quadrant ($+$,) to signify logical addition (conjunction).

22. In his introduction to *Studies in Logic* (p. iv) Peirce remarked on Venn's singular reluctance to modify Boole's system.

23. The conception of duality that follows should be distinguished from that associated with Boole's law of duality, $x(1 - x) = 0$, which is a form of the law of noncontradiction. See Boole's *Laws of Thought* (Dover), pp. 49–51.

24. If we consider the main principles of Boole's calculus (ignoring the formulas that involve subtraction and division), we can list them in a column on the left with their duals (numbered with primes) in a column on the right.

(1)	$xy = yx.$	(1')	$x + y = y + x.$
(2)	$z(x + y) = zx + zy.$	(2')	$z + xy = (z + x)(z + y).$
(3)	If $x = y$, then $zx = zy.$	(3')	If $x = y$, then $z + x = z + y.$
(4)	$x^2 = x$ $[xx = x].$	(4')	$x + x = x.$

Schröder is usually credited with being the first to present the dual laws for logical multiplication and addition in corresponding columns. See his 1877 *Der Operationskreis des Logikkalkuls*. According to Peirce, the practice was probably introduced into geometry by Gergonne about 1820 (CP 4.277).

25. Originally published in the *Zeitschrift für Mathematik und Physik* 25 (1880), 81–94. Republished in translation by Victor H. Dudman in *Southern Journal of Philosophy* 7 (1969), 139–150. In The Johns Hopkins *University Circular* for April 1880, p. 49, Schröder was reported to have made an earlier reference to Peirce's priority in Leo Koenigsberger's *Repertorium der literarischen arbeiten aus dem gebiete der reinen und angewandten mathematik* (Leipzig: B.G. Teubner, 1877–79).

26. See, for example, Marcel Rueff and Max Jeger, *Sets and Boolean Algebra*, Allen and Unwin, London, 1970, p. 21.

27. Arthur N. Prior, in the paper cited in note 2, interprets the nonrelative logic of AJM–1880 as an axiomatic system, although he remarks that Peirce's "metalogic is ... loosely stated, and has something in common with systems of natural deduction" (p. 79). Instead of introducing axioms Peirce identified some fundamental rules of inference, some given as definitions, and a rule (Part 1, §4(2)) for conditionalizing his rules. He did not, however, prove the rule of conditionalization as a theorem, although he did later (in 1898) for his system of Existential Graphs. (See Don D. Roberts, "The Existential Graphs and Natural Deduction," Moore and Robin, pp. 113–114.)

28. CP 3.198ff.

29. The a's in (1) and the c's in (2) and (3) appear as x's in AJM–1880. See note 11 above.

30. The arrangement of terms in (8a) and (8b) differs from the earlier expressions of the law of distribution. From here on I will usually adhere to these forms. By the principles of substitution and commutativity the formulas as expressed earlier can be retrieved.

31. Peirce's note on p. 32 (AJM–1880)—CP 3.199n—reveals that he was familiar with Schröder's *Der Operationskreis des Logikkalkuls* at least by the spring of 1880, and Schröder's reference to Peirce's "On an Improvement in Boole's Calculus of Logic" in his review of Frege's *Begriffsschrift* (see note 25 above) shows that he was acquainted at least with that work of Peirce's by 1880. Records of correspondence in the files of Max Fisch indicate that Peirce and Schröder began their first (about 4-year) period of correspondence around this time.

32. This letter has never been found (nor has Peirce's reply) but these events are described elsewhere in the Peirce Papers.

33. The sign $\prec$ (claw) was introduced by Peirce in 1870 in his "Description of a Notation for the Logic of Relatives...." (CP 3.47; W2.360). Peirce uses the claw as a general connective that can be interpreted as the sign of inclusion, the copula of a categorical assertion, or the connective of a conditional sentence.

34. An abstract of Schröder's paper was printed in the Report of the Fifty-Third Meeting of the British Association for the Advancement of Science, held at Southport in September 1883. (London: John Murray, 1884.) Either Schröder or his printer appear to have transposed the formulas in expressing which can and which cannot be proved. Peirce always identified $a(b + c) \prec ab + ac$ and its dual $(a + b)(a + c) \prec a + bc$ as the contentious formulas. In any case, Schröder announced to the world that Peirce had failed in his 1880 construction. Schröder's full discussion of this matter can be found in his *Vorlesungen über die Algebra der Logik*, Bd. 1, §12.

$a(b + c) = ab + ac$ appears as an axiom in Schröder's 1877 *Der Operationskreis des Logikkalkuls*, (pp. 9–10).

35. Edward V. Huntington, "Sets of Independent Postulates for the Algebra of Logic," *Transactions of the American Mathematical Society* 5 (1904), 288–309. See p. 291n.

36. See MS 547, p. 013, and CP 3.384n (AJM–1885).

37. The independence of $(((a + b) \times c) \prec ((a \times c) + (b \times c)))$ and $(((a + c) \times (b + c)) \prec ((a \times b) + c))$ has been elegantly demonstrated to me by Professor Atwell Turquette, in a letter of 10 June 1985, in which he sketched an independence proof relative to the system that consists of (1)–(3).

38. I will not consider here the extent to which Peirce may have *believed* that (1)–(3) adequately represented the basis of his 1880 logic.

39. CP 3.201. As the first clause of his definition of negation Peirce gave "$\bar{a}$ is of the form $a \prec x$," but he apparently intended x to stand for "what does not occur" (see CP 3.191).

40. In a letter to E.V. Huntington dated 24 December 1903 (MS L210) Peirce wrote about AJM–1880: "You will observe that p. 33 [CP 3.199] follows the whole algebra of the copula, where negation is already introduced & defined."

41. Prior, p. 79. Prior refers to De Morgan's article "Syllogism" in the English Cyclopaedia (1861).

42. John Venn in his *Symbolic Logic*, p. 93 (Chelsea reprint of 1894 second edition),

noted that the rule equating the proposition "xy is a or b" with "$x\bar{a}$ is $\bar{y}$ or b" is "sometimes called Peirce's rule." Venn also gave a more general statement of this rule. I have found no evidence that the name, "Peirce's rule," has been carried forward to denote this, or any other, proposition or rule.

43. In 1898 Peirce gave an explicit statement of "Peirce's Rule" (for his Existential Graphs) which took account of the case where $n = 0$. See my above reference to Don D. Roberts in note 27.

44. Whenever I use the rule of a theorem to justify a step in the proofs that follow, I am abbreviating the proof by omitting a statement of the theorem and an application of MP. See Prior, p. 93, re Peirce's implicit use of the rule that (p and $q \therefore r$) is of the same validity as, and interderivable with (($p \times q) \therefore r$).

45. That is the general purpose of Prior's paper "The Algebra of the Copula." Prior gave (in Polish notation) the following axioms as representing Peirce's 1880 algebra of logic:

1. $(b \prec c) \prec ((a \prec b) \prec (a \prec c))$.
2. $(a \prec (b \prec c)) \prec (b \prec (a \prec c))$.
3. $(a \prec a)$.
4. $(\bar{\bar{a}} \prec a)$.
5. $(0 \prec a)$.
6. $(a \prec c) \prec ((b \prec c) \prec ((a + b) \prec c))$.
7. $((a + b) \prec c) \prec (a \prec c)$.
8. $((a + b) \prec c) \prec (b \prec c)$.
9. $((a \prec b) \prec ((a \prec c) \prec (a \prec (b \times c)))$.
10. $(a \prec (b \prec c)) \prec ((a \times b) \prec c)$.

46. See MS 575, p. 21. MS 575 will be remarked on toward the end of this chapter.

47. Peirce may have begun to question the integrity of his 1880 calculus even before Schröder raised objections. In Note B to *Studies in Logic* (CP 3.328–358), entitled "The Logic of Relatives" and written in the fall or winter of 1882, there is an indication that Peirce was beginning to have some doubts about the derivability of the full law of distribution. He included the problematic part as a definition with what he termed "the main formulae of aggregation and composition," something he had not done in AJM–1880, and characterized the problematic formula as "unobvious and important," while he called the unproblematic part "obvious and trivial." Although in Note B Peirce primarily discusses relative logic he says that relative terms can be aggregated and compounded like others. His remarks about the law of distribution appear to apply equally to both relative and nonrelative logic.

48. According to Peirce (in NEIII-1.289 and CP 2.532) the dilemma was first treated in books on rhetoric but was introduced into logic about 1500 by Laurentius Valla.

49. Peirce would not have claimed that *all* reasoning can be reduced to Barbara, only *deductive* reasoning. (See note 4 above.) I do not know where Whately stood on this point.

50. Whately, *Elements*, 114.

51. A good beginning for understanding the problems of Peirce's logic in the light of the developments following *Principia Mathematica* is Randall R. Dipert's dissertation "Development and Crisis in Late Boolean Logic: the Deductive Logics of Peirce, Jevons, and Schröder" (Indiana University, 1978).

52. MS 527, pp. 13, 15.

53. Peirce claimed (CP 3.384n.) that in Note B to *Studies in Logic* he had "virtually" anticipated Schröder in detecting the inadequacy of AJM–1880 (see note 47 above).

54. *Studies in Logic, By Members of The Johns Hopkins University.* Edited by C.S. Peirce, Boston: Little, Brown, and Company. Reprinted in 1983 by John Benjamins Publishing Company.

55. MS 527, p. 12. Mitchell also used this disjunctive definition of implication. Peirce does not appear to have been familiar at this time with the propositional logic of the Megarian and early Stoic logicians, although he knew of it by way of the Epicurean critiques. His student, Allen Marquand, lists Zeller's 1880 *Stoics, Epicureans, and Sceptics* as a reference in his "The Logic of the Epicureans" which appeared in *Studies in Logic*. Epicurean logic, particularly that of Philodemus, was the subject of Marquand's doctoral thesis for The Johns Hopkins University, for which Peirce was his advisor.

56. See MS S38 (NEIII-1.284ff.), from a correspondence course Peirce offered after he left The Johns Hopkins.

57. "On the Algebra of Logic: A Contribution to the Philosophy of Notation," *The American Journal of Mathematics* 7 (1885) 180–220; CP 3.359–403. Peirce "proved" his fifth icon by a short truth-table method. (See MS S527, pp. 2 and 14, for early appearances of Peirce's law.)

Łukasiewicz may have been the first to refer to the fifth icon as "Peirce's Law." (See J. Łukasiewicz, *Selected Works*, ed. L. Borkowski, Amsterdam: North-Holland Pub. Co. (1970), p. 296.) The importance of "Peirce's law" for the development of modern logic is widely recognized. According to Hao Wang "it is now a familiar result that it is necessary to add Peirce's law ... to render the positive implicational calculus classically complete" (*From Frege to Gödel: A Source Book in Mathematical Logic, 1879–1931*, ed. Jean van Heijenoort, p. 416). A more obscure, though interesting, reference to "Peirce's law" (without reference to Peirce) was made by Bertrand Russell in a letter to Frege (12.12.1904: *Philosophical and Mathematical Correspondence, Gottlob Frege*, eds. Gabriel, Hermes, Kambartel, Thiel, and Veraart, Chicago, University of Chicago Press, 1980, p. 168):

> For negation I use as a primitive law:
> $$\vdash \therefore p \supset q . \supset . p : \supset . p$$
> which is hardly self-evident.

Russell called this formula the "principle of reduction." His remark is curiously similar in manner of presentation to Peirce's own statement in AJM–1885, a paper Russell appears to have been acquainted with.

See Atwell R. Turquette's "Peirce's Icons for Deductive Logic" (*Studies*, pp. 95–108; see note 2 above) for an assessment of AJM–1885. Turquette gives many helpful interpretative suggestions.

58. The Peirce–Huntington correspondence referred to here is all located in MS L210 of the Peirce Papers in the Houghton Library at Harvard University.

59. The drafts of the letter that give the proof appear to deal almost entirely with it and contain no general discussion of Huntington's paper. It would seem, then, that Peirce's "painstaking" letter must have been sent separately from the one which contained the proof.

The following is the proof as it appears in the most finished draft (with only stylistic alterations). The numbers in parentheses refer to formulas in Peirce's published paper (AJM–1880), and those in square brackets are steps in the proof. Peirce retained + (but with a curved horizontal line) as his sign of disjunction but used a raised dot (·) as the sign of conjunction.

Here is the "tedious" proof of the distributive principle to which I referred *Amer. J. Math.* III 33.

$$\text{By (3)} \quad a \cdot c \prec c,$$
$$\text{By (3)} \quad b \cdot c \prec c,$$
[1] $\therefore$ by (2) $a \cdot c + b \cdot c \prec c.$

By (3) $a \cdot c \prec a$ and by (4) $a \prec a + b \therefore a \cdot c \prec a + b,$
By (3) $b \cdot c \prec b$ and by (4) $b \prec a + b \therefore b \cdot c \prec a + b,$
[2] $\therefore$ by (2) $a \cdot c + b \cdot c \prec a + b.$

From [1] and [2] by (2) $a \cdot c + b \cdot c \prec (a + b) \cdot c.$

———————————

Now if $\overline{(a + b) \cdot c} \prec a + b \cdot c$ (that is, if $(a + b) \cdot c \,\overline{\prec}\, a + b \cdot c$)
there must be something, m, such that

[3] $m \prec \overline{(a + b) \cdot c}$

while

[4] $m \prec a + b \cdot c.$

[5] From [3] by (3) $m \prec a + b.$
[6] From [3] by (3) $m \prec c.$

From [4] by p. 28 (19) $a + b \cdot c \prec \bar{m}.$
Whence by p. 33 (3) $a \prec \bar{m},$
[7]   and by p. 28 (19) $m \prec \bar{a},$
and by (3) $b \cdot c \prec \bar{m},$
and by p. 28 (19) $m \prec \overline{b \cdot c}.$

This follows from the Dictum de omni. That is, it is part of the definition of the copula $\prec$ that if $y \prec z$ either $y = 0$ or there is something x such that

$x \prec y$ | If $z = 0$ this is so
$x \prec z$ | even if $y = 0$.

It must be so if $y \prec y$.

Then since m is not 0 (when $m \prec b, m \prec c, m \prec \overline{b \cdot c}$ which is absurd),
by (2) either $m \prec c$ contrary to [6],
or (what we are thus forced to) $m \prec \bar{b}.$

But this being the case there must be something, l, such that
[8] $l \prec m,$
[9] $l \prec \bar{b}.$

Whence by p. 28 (19) $b \prec \bar{l}.$
From [8] and [7] $l \prec \bar{a}.$
[10] Whence by p. 28 (19) $a \prec \bar{l},$
and from [9] and [10] by (2) $a + b \prec \bar{l}.$
[11] Whence by p. 28 (19) $l \prec \overline{a + b}.$
[12] But from [8] and [5] $l \prec a + b.$

And since [11] and [12] cannot both be true the hypothesis that $\overline{(a + b) \cdot c} \prec a + b \cdot c$ is absurd.

[13] We have then $(a + b) \cdot c \prec a + b \cdot c.$
Whence by p. 33 (6) $c \cdot (a + b) \prec (b + a) \cdot c,$
and by [13] $(b + a) \cdot c \prec b + a \cdot c,$
and by p. 33 (6) $b + a \cdot c \prec a \cdot c + b,$
and by p. 33 (5) $(a + b) \cdot c \prec (a + b) \cdot c \cdot c$ and by (6), etc.,
$$(a + b) \cdot c \prec c \cdot (a + b) \cdot c \prec a \cdot c + b \cdot c, \text{ etc.}$$

Peirce added this note:

> I give the above for what it is worth. There may be some fallacy in it. But the principle of introducing m and l is certainly not itself fallacious. At least it does not seem to me so. You observe that p. 33 follows the whole algebra of the copula, where negation is already introduced & defined. But my mind is not now near this subject.

On 31 January 1902 Peirce had recorded an earlier version of this proof in his logic notebook, MS 339(c), p. 437.

60. Huntington replied on 7 February 1904:

> There is no doubt in my mind that the demonstration is perfectly complete, if one grants the necessary assumptions from the "algebra of the copula," namely:
>
> (A) if $a \prec b$ is false, then there is an element x, not 0, such that $x \prec a$ and $x \prec \bar{b}$; and
> (B) if $a \prec b$, then inversely $\bar{b} \prec \bar{a}$.
>
> Of these two assumptions, the first one, (A), is the only one which is essential, as the other can be readily derived from it.

61. Huntington, p. 300n. See CP 3.200n.

62. The proof that Huntington (and later, Lewis) gives for the first part of the law of distribution is nearly identical to that of Peirce (except for notational differences), but Peirce does not receive credit for it. Huntington may have worked out this part of the proof himself, or he may have found it elsewhere. According to Peirce, this principle of logic orignated with Boole and was retained by Jevons (AJM–1880; CP 3.200, n. 3), but there is no indication that either Boole or Jevons gave a demonstration. In 1883 when Schröder announced that only part of the law of distribution is demonstrable (see note 25 above), he indicated that the demonstration followed "the method of Ch. Peirce." Thus it is possible that even the proof of this part of the law of distribution came (at least indirectly) from Peirce.

63. Henry H. Crapo and Don D. Roberts, Abstract of "Peirce Algebras and The Distributivity Scandal." *Journal of Symbolic Logic* 34 (1969), 153–154. Peirce's proof, and the significance of AJM–1880 for the development of lattice theory, is discussed in Crapo's *Lattice Theory*, University of Waterloo, 1966.

64. MS 567 contains what remains of Peirce's note to AJM–1885; a large selection was included in the *Collected Papers*, 3.403A–M. In 3.403An Hartshorne and Weiss remark that these notes were written up for publication. It is probably the paper Peirce intended as the second part of AJM–1885, which Simon Newcomb refused to publish in *The American Journal of Mathematics* because Peirce would not say its subject was mathematics rather than logic.

65. Peirce does not explicitly distinguish between definitions, rules, and axioms.

66. Frege is often credited with the discovery of the modern quantifier, and the 1879 publication of his *Begriffsschrift*, at least 4 years before the publication of a paper by Peirce that might be argued to contain modern quantification theory (Note B in *Studies in Logic*) and 6 years before AJM–1885, is strong *prima facie* evidence for Frege's priority. But Hilary Putnam, in a recent paper about Peirce as a logician ("Peirce the Logician," *Historia Mathematica* 9, no. 3, August, 1982) distinguishes between discovery that is historically first and discovery that is *effective* in the sense that it initiates a line of research. He says that Frege discovered the quantifier in the first sense but that Peirce and his students were its discoverers in the second sense.

67. O.H. Mitchell, "On a New Algebra of Logic," *Studies in Logic*, pp. 72–106. Peirce frequently remarked on the importance and originality of Mitchell's work; see, for example, MS 527 and CP 3.363. Besides its importance for the development of quantification theory, Mitchell's work raised doubts for Peirce about the wisdom of basing his logic on implication.

The First Russell Paradox

Irving H. Anellis

1. Introduction

In 1896, Russell read Arthur Hannequin's (1895) book *Essai Critique sur l'Hypothèse des Atomes dans la Science Contemporaine*, which introduced him to Cantor's work for the first time. Shortly thereafter, in 1897, he read Louis Couturat's (1896) book *De l'Infini Mathématique* and published a review of that book in the journal *Mind* (Russell, 1897). Couturat gave a very careful exposition of Cantorian set theory and, most unusual for its day, a very favorable one. Russell at first rejected Cantorian set theory, for reasons which I detail in Anellis (1984b). Nevertheless, after reading Russell's review of his book, Couturat established a correspondence with Russell (1897–1914) that lasted until Couturat's tragic death. Thanks to Couturat's efforts, Russell came to appreciate better and understand Cantor's work, and soon entered into correspondence with Cantor. Cantor soon furnished Russell with reprints of his work including *Über die elementare Frage der Mannigfaltigkeitslehre* (1892). This new attitude towards Cantorian set theory, however, did nothing to diminish Russell's critical eye. And after reading Cantor (1892), Russell was able to obtain the first version of the Russell paradox. In §3, which is the heart of this chapter, I use unpublished documentary evidence to show that Russell obtained a version of his famed paradox much earlier than has been commonly assumed, and, with the aid of a work of Crossley (1973) in which the Russell paradox was derived directly from the Cantor paradox, I show how Russell had done the same, as early as December 1900. For these reasons, it becomes necessary to sort through the complex and sometimes confused chronological debates on the origin of the Russell paradox (§1) and the equally complex and contentious chronological debate concerning the connections of the Russell paradox with the other set-theoretic paradoxes, especially the Cantor and Burali-Forti paradoxes, with which the Russell paradox is linked in the historical literature (§2).

2. Chronology of Discovery

There is some confusion about the chronology of the discovery of the Russell paradox. J.B. Rosser (1953, p. 201) states that "the Cantor and Burali-Forti paradoxes were discovered about the same time as the Russell paradox...," although he gives no specific year or date. Gregory H. Moore (1982 p. 89) notes that Zermelo, working in Schröder's algebraic logic, obtained "what later became known as Russell's paradox, two years before Russell himself" found the paradox. Since the date of Zermelo's discovery, using Moore's evidence, may be either 1899 or 1900, then we are led to conclude that Russell's discovery occurred in either 1901 or 1902. Most likely, Moore had 1901 in mind, in light of the evidence provided by Jean van Heijenoort (1967a, Editor's Introduction, p. 124) and others.[1] Taking our cue from Rosser (1953, p. 201), let us now examine the history of the Cantor and Burali-Forti paradoxes.

We know that Cantor and Burali-Forti found the paradoxes of the greatest cardinal and the greatest ordinal in the years 1897–1899. The usual date for Cantor is 1899, with reference to his 28 July 1899 letter to Dedekind (Zermelo, 1932), while others refer to Cantor's (1895–97) *Beiträge zur Begründung der transfiniten Mengenlehre*. The reference for Burali-Forti (1897) is his *Una questione sui numeri transfiniti*. The discussion of the Cantor and Burali-Forti paradoxes is significant in any consideration of the discovery of the Russell paradox, thanks to the connections which Russell's paradox is said to have to them. (We shall examine these connections in §2.)

The usual date given for the discovery by Russell of his paradox is based on Russell's letter to Frege of 16 June 1902.[2]

We must note that Russell (1901, p. 95) alludes to the Cantor paradox in his survey *Recent Work on the Principles of Mathematics*. Jourdain (1912, p. 153) refers to this article; but there is no explanation of the Cantor paradox, and no mention at all of the Russell paradox in that survey. Beyond that, we can refer to Ivor Grattan-Guinness (1978, p. 135), who in turn refers to Philip E.B. Jourdain (1913, p. 146), where, according to Grattan-Guinness, Jourdain claims that Russell discovered Cantor's paradox in January 1901, and his own in June of that same year. If we refer to Jourdain's (1915, p. 206) notes in his translation of Cantor (1895–97), then again we are given the year 1901, but without reference to the month, and asked to refer to Jourdain (1912). Jourdain (1913, p. 146) does refer to a note sent to him by Russell, according to which Russell corrects Jourdain's (1912, p. 153) assertion of the 1912 article that the survey (Russell, 1901) was written in January 1901 and that he "found his contradiction in June. In January he had only found that there must be *something* wrong..." (Jourdain, 1913 p. 146), without knowing what. It worried Russell that "Cantor's proof would show that there is a greater transfinite cardinal than the number of all entities." Jourdain is not more explicit than that. In his *Introduction to Mathematical Philosophy*, Russell (1919, p. 136) also recalls 1901 as the year in which he discovered his paradox, while attempting "to discover some flaw in Cantor's proof that there is no greatest cardinal."

Given the conflicting claims made by Russell himself, both to Jourdain and others, as Grattan-Guinness (1978, pp. 134–135) delineates, it is perhaps best to rely upon archival materials rather than on Russell's retrospective recollections. This becomes all the more important inasmuch as, in the only reference to the Russell paradox in the published Russell–Jourdain correspondence (Grattan-Guinness, 1977, pp. 24–29), no mention is made of the chronology of the discovery.

A.R. Garciadiego Dantan (1983) in his thesis has also considered the history of the Russell paradox.[3] In the abstract, Garciadiego (1984) specifically mentions correspondence between Russell and Couturat. As I shall show (§3), the letter of 8 December 1900 from Russell to Couturat in the Russell Archives conveys the *first version* of the Russell paradox. If this is the case, Bunn's claims (1980, pp. 238–239) for January 1901 would be substantiated. But despite reference to the Russell–Couturat correspondence, Garciadiego gives the date of May 1901, referring to the penultimate draft of Russell's *Principles of Mathematics*, as the time of Russell's first *statement* of his paradox.

Except for Blackwell, Garciadiego Dantan, Grattan-Guinness, Griffin, and Moore, none of the people who contributed to the discussion of the chronology of the discovery of the Russell paradox (apparently) had access to the Russell Archives, and so they based their assertions on available published evidence, or on Russell's own, sometimes conflicting, recollections (see Grattan-Guinness, 1978, pp. 134–135, ftn. 4). Of these people, none cite Russell's 8 December 1900 letter as the source for the Russell discovery. Therefore, the scholarly consensus among Russell experts who have had access to the Russell Archives assigns June 1901 as the earliest time for discovery by Russell of his paradox. Of these, only Blackwell and Grattan-Guinness have seriously considered the chronology, and Blackwell (1984–5, p. 276) did so only in a passing reference.[4]

Before leaving this review of the literature on the chronology of the discovery of the Russell paradox for a consideration of purely technical questions on the connection between the Russell paradox and the Cantor and Burali-Forti paradoxes, and the correlation of these questions with the chronological question, it would not be amiss to mention Russell's very close collaborations with G.H. Hardy and A.N. Whitehead during the period from the 1890s to 1913, and their mutually shared interests in the paradoxes of set theory. But there is nothing extant in the Russell Archives, unfortunately, in the way of notes or correspondence between Russell and Hardy or Russell and Whitehead, which can further aid our study of this matter.

3. The Connection of the Russell Paradox with the Cantor and Burali-Forti Paradoxes, and the Chronology of the Russell Paradox

There are, in the literature, five distinct theories of how Russell discovered his paradox, each with its own variations. The first holds that Russell obtained his paradox by deriving it from the Burali-Forti paradox (Temple, 1981, p. 261). Another holds that it was obtained by an analysis of the proof of the Burali-Forti paradox (Peano, 1902–1906, p. 144). A third asserts that the Russell paradox is derived from the Cantor paradox. It is espoused by Kleene (1971, p. 37); by Beth (1966, p. 484), who fudges by saying that after Cantor's paradox became known to Russell the latter constructed his own paradox; by Crossley (1973) and Bunn (1980, p. 239), who show how to obtain the Russell paradox from the Cantor paradox; by Grattan-Guinness (1978, p. 130), who cites Crossely and Bunn; and by H.C. Doets (1983, p. 300), who goes so far as to claim that Russell's paradox *is* Cantor's paradox, and whose reconstruction of the Russell paradox from the Cantor paradox is most inelegant compared to Crossley's. Another theory asserts that Russell obtained his paradox from an analysis of Cantor's paradox and the proof of the Cantor theorem. It is espoused by Irving M. Copi (1971, p. 7); P.H. Nidditch (1962, p. 70); A. Levy (1979, p. 87), who gives a reconstruction; by N.B. Cocchiarella (1974, p. 553), who states that the argument for the Russell paradox is a variant of the argument for Cantor's theorem; by Grattan-Guinness (1978, p. 129), referring in particular to Russell (1903, pp. 366–367); and by Grattan-Guinness (1980, p. 76). This is the theory that Russell himself adopted in his *Introduction to Mathematical Philosophy*. The fifth theory is that the set-theoretic paradoxes share a common structure, which is exposed by the Russsell paradox. According to this theory, the Cantor and Burali-Forti paradoxes are variants or special cases of the Russell paradox. Its advocates are Copi (1971, p. 7), van Heijenoort (1967b, p. 46), Temple (1981, p. 33), and Rosser (1953, p. 205). The most common version of this theory is, from the philosophical perspective, that the paradoxes share the structure of the "vicious circle principle," as defined by Chihara (1973, p. 3) and Poincaré (1906, pp. 307–308).

Grattan-Guinness does not make clear what relation, if any, there is between the Russell paradox and the paradoxes of the greatest cardinal and the greatest ordinal. He espouses at once both the third and fourth theories. J. van Heijenoort (1967b, p. 46) makes it clear that the paradox of the greatest ordinal is the Burali-Forti paradox, and the paradox of the greatest cardinal is the Cantor paradox, and is due to Cantor himself, as a direct result of the power set axiom—or Cantor theorem, van Heijenoort also makes it clear that the Cantor and Burali-Forti paradoxes arise from technical difficulties in Cantor's naive set theory, while the Russell paradox, which is of the *same logical kind*, is more basic insofar as it involves "the bare notions of set and element" (van Heijenoort, 1967a, p. 124, Editor's Introduction; 1967b, p. 46). Rosser (1953, p. 205) notes that the Russell paradox proves that $\sim(x \in x)$ does not determine a class, whereas the Cantor and Burali-Forti paradoxes prove that additional statements (about largest cardinal and largest ordinal) do not determine classes. Thus, it is not surprising that van Heijenoort should conclude that part of the stimulus for Russell's discovery of his paradox was a reflection about the Cantor and Burali-Forti paradoxes and their *structure*. Grattan-Guinness (1972), however, quotes Russell's 15 March 1906 letter to

P.E.B. Jourdain, to the effect that Russell engaged in serious reflection about the Burali-Forti paradox only as late as 1905. Despite Grattan-Guinness's inferences, there is *no* explicit statement in Russell's letter to Jourdain linking the Burali-Forti paradox to the Russell paradox. While Russell must have been mistaken in his recollection of 1905 as the year in which he studied the Burali-Forti paradox, given his references to that paradox in *The Principles of Mathematics*, p. 323, of 1903, we are led nevertheless to connect the chronology of the Russell paradox to the Cantor paradox alone, as evidence provided by Coffa (1979) (which we shall consider momentarily, in §3) suggests. This does nothing, however, to invalidate the theory that the Burali-Forti and Cantor paradoxes share the logical structure of the Russell paradox—the most basic paradox of set theory—and that the Burali-Forti and Cantor paradoxes are just variants or special cases of the Russell paradox.

As is well known, the Russell paradox, in its present form, stems from Frege's allowing a function to be taken as an indeterminate argument for another, higher-order function. Thus, as van Heijenoort points out (1967b, p. 46), it is equivalent in structure to the Burali-Forti and Cantor paradoxes. Let S be the set of all sets, having cardinality α. If the sets composing the set S are well-ordered aggregates, that is, have each a least element and a well-defined linear succession of elements, then we ask whether S itself is well-ordered. Now let U be the set of all subsets of S. Then the cardinality of U must be greater than α. Now according to Cantor (1883a, p. 169), U must also be well-ordered. For Russell in the *Principles*, it does not follow that if each subset of U is well-ordered, then so is U. Thus, for Burali-Forti and Cantor, the question is whether the aggregates defined by the ordinal of U and the cardinal β of U are well-ordered.

Now we recall that the Russell set r is the set whose elements are sets which are not members of themselves. For all x, $x \in r$ if and only if $x \notin x$ (which Rosser (1953, p. 205) wrote as $\sim(x \in x)$); we get immediately that $r \in r$ if and only if $r \notin r$ by substitution using universal instantiation. As Russell noted, the problem becomes that of whether r can count as a class; for if x is the class of all classes, then necessarily $r \in x$ if r is a class. Thus, the problem now is whether the class of all classes is a class. At first, Russell thought it was. But this is exactly the problem raised by Russell when first examining Cantor, and it is raised, as we shall see (in §3), in exactly these same terms in Russell's 8 December 1900 letter to Couturat. Thus, Russell is essentially raising the question of whether x is a well-ordered aggregate if we can obtain $r \in r \Leftrightarrow r \notin r$ from $x \in r \Leftrightarrow x \notin x$. For Russell it does not follow from $x \in r$ and x's being well-ordered that r is a well-ordered aggregate. Moreover, if x is the class of all classes, it becomes an open question whether x is well-ordered. But if in fact the class of classes is a class, then the set-theoretic paradoxes arise. Thus, Rosser (1953, p. 23) takes the Russell paradox to warrant a distinction between sets and classes, according to which all sets are classes while some classes are nonsets.[5] In 1900, however, Russell made no such distinction between sets and classes. He held that the class of classes is a class. Thus, the set-theoretic paradoxes were built into Cantor's naive set theory.

4. The First Russell Paradox

The first Russell paradox, or more carefully, the first version of the Russell paradox, is connected with a mistake which Russell discovered in Cantor's (1892) proof of the nonexistence of a greatest cardinal, i.e., precisely in connection with the Cantor paradox. The Cantor paradox arises in Cantor's proof of his theorem that there is no greatest cardinal. Thus those who connected the Russell paradox to the Cantor paradox were correct; they were mistaken only in believing that Russell discovered his paradox in 1901.[6] The first Russell paradox was stated by Russell in a letter to Couturat of 8 December 1900. According to this letter, Cantor's attempt to prove the inexistence of a greatest cardinal contains a mistake which in fact, thought Russell, proves exactly the opposite of what Cantor intended. The subject of this proof is now known as Cantor's theorem, which reads (Cantor, 1892, p. 77): "... *die Mächtigkeiten wohldefinierter Mannigfaltigkeiten kein Maximum haben oder, was dasselbe ist, ... jeder gegebenen Mannigfaltigkeit L eine andere M an die Seite gestellt werden kann, welche von stärker Mächtigkeit ist als L*," ("the power of a well-defined set has no maximum or, what is the same, for each given set L another M can be found whose power is greater than that of L.")

Cantor gave two proofs for his theorem. The first is the famous diagonal argument which, for any given set, allows us to construct a new set from the elements of the original set such that the new set contains some elements not contained in the original set, so that the cardinality of the new set is greater than that of the original set. (This diagonal proof is carefully and fully explained by Joseph W. Dauben (1980, pp. 204–205), and is hastily summarized by Grattan-Guinness (1978, pp. 127–128).) Our concern is with the second proof. The first proof shows that the cardinality of the new set is greater than that of the orignial set. The second proof purports to show that the new set is neither smaller than nor equal in cardinality to the original set. This second proof runs as follows.

Let L be the linear continuum, let R be the set of all reals x in the interval $[0, 1]$, and let M be the set of all one-to-one functions $f(x)$ from elements $x \in [0, 1]$ onto the set $\{0, 1\}$. Now if there is an isomorphism $f : x \to \{0, 1\}$ such that $f(x_0) = 1$ and $f(x_v) = 0$ (for all $v \neq 0$, $x_v \in [0, 1]$), then M does not have a smaller cardinality than L. Now suppose that card$(M) =$ card(R). Then there is a binary function $F(x, z)$ from f into elements $z \in [0, 1]$, and F is an isomorphism. It is here, at this stage in his proof, that Cantor announces that a contradiction follows. Define $g(x)$ such that

$$g(x) = \begin{cases} 1 & \text{if} \quad F(x, x) = 0, \\ 0 & \text{if} \quad F(x, x) = 1, \end{cases}$$

Since $g(x)$ is characteristic over a subset of R, then $g(x) \subset M$. But now $g(x)$ also cannot have a value $z = z_0$ result from $F(x, z)$, since $F(z_0, z_0)$ is not the same as $g(z_0)$. Thus, contrary to hypothesis, no such isomorphism as f can occur, since clearly $f(x_\alpha) = f(x_\beta)$ if and only if $x_\alpha = x_\beta$, by definition of isomorphism. But it does not always necessarily happen that $x_\alpha = x_\beta$. This means that for $f: PA \to A$ (writing PA as the power set of A), f cannot be an isomorphism, since $\text{card}(PA) > \text{card}(A)$. Thus, $\text{card}(M) > \text{card}(R)$. But by hypothesis, $\text{card}(M) = \text{card}(R)$. This contradiction, the Cantor paradox, is a paradox of cardinality.

We may reformulate it in more familiar terms, set-theoretically rather than function-theoretically. Define the power set of a, $P(a)$, as $\{x: x \subset a\}$. Given the power set axiom (i.e., Cantor's theorem), which means that

$$\exists x \forall y (x \in y \Leftrightarrow \forall z(z \in x \Rightarrow z \in a)),$$

write $\forall z(z \in x \Rightarrow z \in a)$ as $x \subset a$. We thus get $P(a) = \exists x \forall y(x \in y \Leftrightarrow x \subset a)$, which means that if a is an element of the set A, it is also an element of the power set of A, and, more importantly, that for any set A, clearly $PA \not\cong A$. Let B be so defined that $B \subset PA \Rightarrow \text{card}(B) < \text{card}(PA)$. Clearly PA contains subsets of A, among these B, including $\varnothing$ and A itself. So $\text{card}(PA) > \text{card}(A)$. In fact, Cantor's theorem says, in modern terminology, that if $\text{card}(A) = \alpha$, then $\text{card}(PA) = 2^\alpha$. But now let S be the set of all sets. Then $\text{card}(PS) > \text{card}(S)$, by the previous result. But by definition of S as the set of all sets, clearly $S \cong PS$. Therefore, we have

$$\text{card}(PS) > \text{card}(S) \ \& \ S \cong PS.$$

This is again the Cantor paradox, and is clearly a paradox of cardinality. But it has now been expressed in Russellian terms, by defining S as the set of all sets (or, what was the same for Russell, the class of all classes). In fact, as we shall now see, Russell in 1900 was already expressing the Cantor paradox in exactly these terms, as the class of all classes being a class. In doing so, he was giving the *first version of the Russell paradox*, which we shall distinguish from the classical version of the Russell paradox found in June 1901, some 6 months later and carried out in the classical version in terms of the Russell set.

Russell wrote, in his letter to Couturat (1900 Russell to Couturat, 8 December 1900);

> I have found a mistake in Cantor, who holds that there is no greatest cardinal number. But the number of classes is the greatest number. The best proof to the contrary that Cantor gives is found in *Jahresber. Deutsch. Math.-Verein.* I, 1892, pp. 75–78. It consists basically, in showing that, if u is a class of which the number is α, the number of classes contained in u (which is 2^α) is larger than α. But the proof presupposes that there are some classes contained in u which are not individuals of u; now if $u = $ Class, this is false: every class of classes is a class.[7]

Unfortunately, Russell did not go on to develop his argument more exhaustively, or provide a formal proof, in his letter to Couturat. Nor can anything be located in the Russell Archives which develops the argument suggested to Couturat in the quoted letter. Nor has Grattan-Guinness, who has done research at the Russell Archives and who has considered the chronology of the origin of the Russell paradox, given any indication of having found evidence of an early (prior to June 1901) version of the Russell paradox. He does give (Grattan-Guinness, 1978, p. 127), as we have noted, a reference to Cantor's (1892) paper, including the passages to which Russell referred in his letter to Couturat. But Grattan-Guinness (1978, pp. 127–128) is concerned here exclusively with the development of the diagonal argument which Cantor used to prove his theorem and prove the nondenumerability of the linear continuum. Grattan-Guinness (1978, pp. 129–130) attempts to show that the Russell paradox can be obtained from the Cantor paradox within the argument for Cantor's theorem, but does not conclude that Russell actually did so. Instead he discusses how a reconstruction of the Russell paradox from Cantor's argument *might* go, relying on that alluded to by Bunn (1980, p. 239) and carried out by Crossley (1973) to show that such a reconstruction is possible. Grattan-Guinness (1978, p. 128) incorrectly asserts that Cantor failed to associate M with the linear continuum, by failing to assign elements of $x \in [0, 1]$ to $\{0, 1\}$. But as we have seen, this is precisely what Cantor (1892, p. 77) *did* do, by allowing $f(x)$ to be a subset of M taking $x_0 \in [0, 1]$ to 1 and every other $x \in [0, 1]$ to 0, while $L = x \in [0, 1]$. Elsewhere, Cantor had already shown that there is an isomorphic correspondence between the points in the topology of L and the elements of the set of reals (see, e.g., Cantor (1883a, p. 343)).

The late J. Alberto Coffa does much better. His 1979 article (Coffa, 1979 p. 33) quotes Russell's 8 December 1900 letter to Couturat and claims (p. 37) that, thinking back to that letter, Russell, around May 1901, realized that the thoughts expressed to Couturat were merely—in Coffa's words—a "gestalt switch away from the celebrated paradox." Moore (1988, p. 52) argued that Russell in his 8 December 1900 letter to Couturat "expressed the paradox of the largest cardinal..., but still did not regard it as a paradox." Moore (1988, p. 53) goes on to say that it was not until May of 1901, however, "that Russell discovered Russell's paradox *as a paradox* ...". I go further than Coffa and Moore.

It is evident that Russell, although he did not work through the details, declaring to Couturat that it would have been too involved to do so, did in fact obtain a *primitive version of the Russell paradox*, that is, the *first version of the Russell paradox*, on 8 December 1900. He did so by restating, in set-theoretic terms, the Cantor paradox, that is, by substituting the class of classes for the power set PS and substituting the class contained in the class of classes for the set S. Russell was no longer considering the cardinality of S and of PS, but *class inclusion*. It was left only to rewrite this in terms of the Russell set, and easy step, and to consider the nature of class inclusion with respect to the Russell set.[8] This may also be considered in terms of the well-ordering of the set of all sets as much as in terms of self-reference. What Coffa meant by a "gestalt switch" enabling Russell to see the classical version of his paradox embedded in the analysis of the Cantor paradox, I suggest, is appropriately understood in relation to Russell's statement to Jourdain, reported in Jourdain (1913, p. 146), of the realization that something was wrong with Cantor's proof of his theorem, without knowing what. It is appropriate to interpret this to mean that Russell recognized the problem, and even what the problem was, without yet being able to full apprehend its portent for set theory and without yet having the requisite tools to attempt a solution.

Despite these uncertainties, we can see that the core of the classical Russell paradox is already present in the letter of 8 December 1900 by following the work of J.N. Crossley (1973). He was able, without having access to the letter, to directly translate into set-theoretic terms Cantor's function-theoretic treatment of the cardinality of the power set as given in Cantor (1892), and thereby he independently obtained a version of the Russell paradox.

Crossley (1973) show that the cardinality of PS and of S are not equal, which is half of Cantor's argument: namely that $\text{card}(PS) > \text{card}(S)$. Take S to be *the universe*, and thus obtain $\text{card}(PS) = \text{card}(S)$, i.e., $PS \cong S$. If f is an isomorphism between PS and S, then for $y = \{x \in S : \sim(x \in f(x))\}$, we obtain $y \in S \Rightarrow y \in PS$. This implies that for some $z \in S$, $y = f(z)$. Then for all $x \in S$, we get $x \in y \Leftrightarrow \sim(x \in f(x))$, which means that $x \in f(z) \Leftrightarrow \sim(x \in f(x))$. If $x = z$, then by substitution, we immediately get $z \in f(x) \Leftrightarrow \sim(z \in (f(z))$, which is the classical Russell paradox. It also follows that there can be no such isomorphism f between S and PS. Moreover, if we replace PS with the class of classes, S with the class u in Russell's letter to Couturat, and substitute the nonisomorphism of $PS > S$ with the condition of being the Russell set, we obtain the standard version of the classical Russell paradox. It is sufficient to point to the class of classes as being a class—introduced in the letter to Couturat—together with the assertion that there are some classes in $u =$ which are not individuals of u and letting $u =$ class, to obtain the first, or primitive, version of the Russell paradox from the Cantor paradox as presented Cantor (1892). Most assuredly, Russell is already interested, in this letter, in the paradox involving set inclusion, and not the cardinalities of the sets. Thus, Crossley's derivation of the Russell paradox from the Cantor paradox as presented in 1892 confirms my conjecture (Anellis, 1983, 1984a, b, pp. 9–10, 1987, pp. 20–25) that Russell's criticism of Cantor is the unproved anticipation, or primitive version, of the Russell paradox, and predates by approximately 18 months the official classical version, the final version presented in his letter of Frege. At most, we have the *first version of the Russell paradox*, obtained some 6 months before the first version of the classical Russell paradox. What is patently missing, however, in Russell's appraisal of Cantor's argument in the letter to Couturat, is the "very long argument" that Russell mentioned to Couturat which would justify Russell's appraisal. Only that would permit the historian to confirm that Russell actually came to a *realization* that he had the paradox in hand when he wrote the letter, or to determine how close he came there to fully worked out precursor of the classical Russell paradox.

Acknowledgments

J.N. Crossely informed me about his derivation of the Russell paradox from the proof of Cantor's theorem in a note dated 15 February 1984. Kenneth Blackwell of the Russell Archives at McMaster University and Nick Griffin [formerly] of the Bertrand Russell Editorial Project at McMaster University expertly and expeditiously provided answers to my inquiries, and copies of required documents. Their assistance is greatly appreciated. Thomas Drucker's expert use of the editor's blue pencil went a long way towards improving what began as a long and rambling essay, and has earned him the gratitude of readers of this chapter. I also wish to thank Alexander Abian for his careful reading.

Notes

1. In the television series 'Nova' program "A Mathematical Mystery Tour," prepared by BBC-TV in 1984 and broadcast on 5 March 1985. Moore explicitly refers to the traditional date of June 1901, in connection with the letter to Frege of 16 June 1902. (See the transcript of this program, p. 12; transcript #1208, WGBH, c. 1985.)

2. Kleene (1967, p. 186) gives the year as 1902, as do the historians of mathematics Carl Boyer (1968, p. 663) and George Temple (1981, p. 32), and the historians of logic William and Martha Kneale (1962, p. 652). Elsewhere, Temple (p. 261) considers June 1901 as the discovery month, in connection with Russell's (1903) writing of the *Principles of Mathematics* (which was undertaken in 1900) and with his study of the Burali-Forti paradox. Temple there gives the letter to Frege as Russell's first *announcement* of the paradox. Guiseppe Peano (1902–1906, p. 144), Evert W. Beth (1966, p. 207), Azriel Levy (1937, p. 6), and Kleene (1971, p. 37) give the year as 1903, which, as Grattan-Guinness (1980, p. 76) and R. Bunn (1980, p. 238) note, is actually the year of first *publication* of the Russell paradox (in *Principles of Mathematics*, p. 323) and Frege's *Grundgesetze der Arithmetik*, (1903, II, pp. 253–265). P.H. Nidditch (1962, p. 70), H. DeLong (1970, p. 81), and Jean van Hiejenoort (1967a, Editor's Introduction, p. 124), in their histories of logic, give the year 1901 for Russell's discovery of his paradox, as does the chief Russell Archivist, Kenneth Blackwell (1984–5, p. 276), who cites for his evidence the penultimate draft of the *Principles of Mathematics* and Russell's own claims (see Russell (1919, p. 136)). Blackwell and van Heijenoort give the date as June 1901, Blackwell citing archival evidence. R. Bunn (1980, pp. 238–239) gives January 1901 as the proximate date of Russell's discovery, however, and 1901 as the year of his first mention of it in print, although there is little evidence adduced by Bunn to support his conclusion.

3. By his rather strange and elaborate subtitle to his thesis—Bertrand Russell and the origin of the set-theoretic paradoxes: An inquiry into the causes that allowed and motivatd him to create some of the best-known paradoxes and how others were discovered—Garciadiego appears to suggest that Russell discovered the Cantor and Burali-Forti paradoxes prior to their discoveries by Cantor and Burali-Forti. Moreover, Moore and Garciadiego's article (1981) claims explicitly that Russell discovered the Burali-Forti paradox before Burali-Forti himself did so. This is clearly impossible, since Burali-Forti published his paradox in 1897, and it is based on his reading of Cantor's 1892 and 1895 papers while Cantor discovered his paradox not later than 1899, and by many accounts in 1897, and in fact one can find it as early as 1892. In any of these cases, Russell could not possibly have discovered the Cantor or Burali-Forti paradoxes earlier than 1897; as his notebook, *What Shall I Read?* (Russell, 1983), indicates, Russell did not begin reading Cantor in a serious and systematic way until the latter part of 1897, after completing his review of Couturat. Since I have not had

direct access to Garciadiego's work, I shall disregard it for this study. I have seen the abstract and have also benefitted from discussions about it with Kenneth Blackwell of the Russell Archives and Nick Griffin (formerly) of the Bertrand Russell Editorial Project.

4. If we turn to Grattan-Guinness, we find that our confusion concerning this chronology remains undiminished. Thus, Grattan-Guinness (1978, p. 129) writes that "Cantor was the first mathematician to discover paradoxes in set theory. Around the time of Russell's discovery of his work he found the paradoxes of the greatest ordinal and the greatest cardinal. ... Russell discovered the cardinal paradox for himself early in 1901, and sometime in the next 6 months found his own." It is not immediately clear, from Grattan-Guinness's equivocal use of pronouns, that the one who discovered the paradoxes of the greatest ordinal and the greatest cardinal, after Russell discovered Cantor's work, was Cantor, not Russell; but one must assume from the first sentence quoted that Grattan-Guinness meant that Cantor discovered the Cantor paradox (and the Burali-Forti paradox) about the time that Russell discovered Cantor's work. We must also immediately point out what is either a very distinct error or, minimally, the source of additional confusion; Grattan-Guinness *seems* here to suggest that Russell discovered Cantor's work in or around 1901, and as a result found the Cantor paradox in early 1901, whereas I have already shown—and as Grattan-Guinness himself is in fact aware—that Russell first became aware of Cantor's work as early as 1896, at which time he began to read French translations of Cantor's work published in *Acta Mathematica* of 1883, as I pointed out (Anellis 1983, 1984b, 1987, pp. 1–4) (see Anellis 1984a, p. 2 for a list of Cantor's papers read by Russell in 1896, as based on Russell's own records in *What Shall I Read?*), while his first favorable contact with Cantor's work was made in connection with Couturat's book (1896), that is, in 1897, the very year in which Burali-Forti published his paradox and the very year in which Cantor published Part II of Cantor (1895–97), in which is evident the Cantor paradox.

5. In same way, Cantor (1895–97) himself, in consideration of his own paradox, eventually came to distinguish between complete and incomplete multiplicities.

Whether such distinctions are sufficient to avoid the paradoxes has been rendered questionable by Alexander Abian, who has formulated the paradox $A \in x \Leftrightarrow x \notin x$, allowing us to obtain $A \in A \Leftrightarrow A \notin A$. On this question, Abian states (private communication, 20 December 1985):

Another example which shows that in general the assertion of the existence of sets which are supposed to satisfy some unrestricted property is not permissible and may lead to a paradox, is the following assertion of the existence of a set A via:

(1) $(x)(A \in x \Leftrightarrow x \notin x)$

or which is the same as:

$(x)(A \notin x \Leftrightarrow x \in x)$.

Note that neither is equivalent to Russell's example, i.e.,

$(x)(x \in A \Leftrightarrow x \notin x)$

or

$(x)(x \notin A \Leftrightarrow x(-x)$.

Considering (1), we see that A can neither a set nor a class because replacing x in (1) by the empty see $\emptyset$ we have:

$A \in \emptyset \Leftrightarrow \emptyset \notin \emptyset$, where $A \in \emptyset$ is always false (regardless of whether A is a set or a class) and $\emptyset \notin \emptyset$ is always true, which gives us False $\Leftrightarrow$ True—which is a paradox.

The conclusion from Abian's paradox is that neither sets nor classes should be formulated in terms of arbitrary and unrestricted (that is noncanonical) properties and that some axioms prescribing some rules for formations of sets and classes should exist which hopefully do not yield paradoxes (and they should be appropriatedly corrected in case they yield paradoxes). The first published announcement of the Abian paradox was published by Anellis (1988).

6. Russell himself is responsible for popularization of this error. In Russell (1919, p. 136) he wrote, "When I first came upon this contradiction, in the year 1901. I attempted to discover some flaw in Cantor's proof that there is no greatest cardinal...." Thus, his memory is faulty on the date, but correct on the circumstance of the connection of the Russell paradox with the Cantor paradox.

7. The original runs, "J'ai découvert une erreur dans Cantor, qui soutient qu'il n'y a pas un nombre cardinal maximum. Or le nombre des classes est le nombre maximum. La meilleure des preuves du contraire que donne Cantor se trouve dans *Jahresber. Deutsch. Math.-Verein. I*, 1892, pp. 75–78. Elle consiste au fond à montrer que, si u est un [sic] classe dont le nombre est α, le nombre des classes contenues dans u (qui est 2^{α}) est plus grand que α. Mais la preuve présuppose qu'il y a des classes contenues dans u qui ne sont pas des individus d'u; or si $u =$ classe, ceci est faux: tout [sic] classe de classes est une classe."

8. In this way, we obtain directly the Grelling paradox of heterologicality (see van Heijenoort (1967b, p. 47)), which is, like the classical version of the Russell paradox, a paradox of self-reference.

References

AR is used to designate unpublished documents located in the Bertrand Russell Archives, Mills Memorial Library, at McMaster University. BREP is used to designate typeset documents prepared by the Bertrand Russell Editorial Project at McMaster University for planned publication in *Collected Papers of Bertrand Russell*; numerical designations refer to tentative placement of these papers by volume number and article number.

Anellis, I.H. (1983), Bertrand Russell's earliest reactions to Cantorian set theory, Preliminary report, *Abstracts Amer. Math. Soc.* **4**, 531.

Anellis, I.H. (1984a), Russell on Cantor: An early version of the Russell paradox? Preliminary report, *Abstracts Amer. Math. Soc.* **5**, 128.

Anellis, I.H. (1984b), Bertrand Russell's earliest reactions to Cantorian set theory, 1896–1900, in J.E. Baumgartner, D.A. Martin, and S. Shelah, (eds.), *Axiomatic Set Theory, Contemp. Math.* **31**, 1–11.

Anellis, I.H. (1987), Russell's earliest interpretation of Cantorian set theory, 1896–1900, *Philosophia Mathematica* (n.s.) **2**, 1–31.

Anellis, I.H. (1988), The Abian Paradox, *Abstracts of papers presented to the American Mathematical Society* **9**, 6.

Beth, E.W. (1966), *The Foundations of Mathematics*, rev. ed., Harper & Row, New York.

Blackwell, K. (1984–5), *The Text of the Principles of Mathematics*, Russell (2), 4 (1984–1985), 271–288.

Boyer, C.B. (1968), *A History of Mathematics*, Wiley, New York.

Bunn, R. (1980), Developments in the foundations of mathematics, 1870–1910, in I, Grattan-Guinness (ed.), *From the Calculus to Set Theory, 1630–1910: An Introductory History*, Duckworth, London, pp. 221–255.

Burali-Forti, C. (1897), Una questione sui numeri transfiniti, *Rend. Circ. Mat. Palermo* **11**, 154–164. (English translation in van Heijenoort (1976a, pp. 104–111).

Cantor, G. (1883a), Essais mathématiques, *Acta Math.* **2**, 305–414.

Cantor, G. (1883b), Grundlagen einer allgemeinen Mannigfaltigkeitslehre *Math. Ann.* **21**, 545–586. (Reprinted in Zermelo (1932, pp. 165–204).)

Cantor, G. (1892), Über die elementare Frage der Mannigfaltigkeitslehre, *Jahresber. Deutsch. Math.-Verein.* **1**, 75–78.

Cantor, G. *Cantor an Dedekind, 28 Juli, 1899*, Zermelo (1932, pp. 443–447). (English translation in van Heijenoort (1967a, pp. 113–117).)

Cantor, G. (1895–97), Beiträge zur Begründung der transfiniten Mengenlehre, *Math. Ann.* **46**, 481–512; **49**, 207–246. (English translation in Jourdain (1915).)

Chihara, C.S. (1973), *Ontology and the Vicious Circle Principle*, Cornell University Press, New York.

Cocchiarella, N.B. (1974), Fregean semantics for a realist ontology, *Notre Dame J. Formal Logic* **15**, 552–568.

Coffa, J.A. (1979), The humble origins of Russell's paradox, *Russell* (o.s.) nos. 33–34, 31–37.

Copi, I.M. (1971), *The Theory of Logical Types*, Routledge & Kegan Paul, London.

Couturat, L. (1896), *De l'Infini Mathématique*, Félix Alcan, Paris.

Crossley, J.N. (1973), A note on Cantor's theorem and Russell's paradox, *Austral. J. Philos.* **51**, 70–71.

Dauben, J.W. (1980), The development of Cantorian set theory, in I. Grattan-Guinness (ed.), *From the Calculus to Set Theory, 1630–1910: An Introductory History*, Duckworth, London, pp. 181–219.

DeLong, H. (1970), *A Profile of Mathematical Logic*, Addison-Wesley, Reading, MA.

Doets, H.C. (1983), Cantor's paradise. *Nieuw Arch. Wisk.* **1**, 290–344.

Frege, G. (1903), *Grundgesetze der Arithmetik II*, Pohle, Jena.

Garciadiego Dantan, A.R. (1983), *Bertrand Russell and the Origin of the Set-Theoretic Paradoxes*, Ph.D. Thesis, University of Toronto.

Garciadiego Dantan, A.R. (1984), Abstract: Bertrand Russell and the origin of the set-theoretic paradoxes, *Dissert. Abstracts International* **44**, 2864-A.

Grattan-Guinness, I. (1972), Bertrand Russell on his paradox and the multiplicative axiom: an unpublished letter to Philip Jourdain, *J. Philos. Logic* **1**, 103–110.

Grattan-Guinness, I. (ed.) (1977), *Dear Russell—Dear Jourdain: A Commentary on Russell's Logic, Based on His Correspondence with Philip Jourdain*, Columbia University Press, New York.

Grattan-Guinness, I. (1978), How Bertrand Russell discovered his paradox, *Historia Math.* **5**, 127–137.

Grattan-Guinness, I. (1980), Georg Cantor's influence on Bertrand Russell, *Hist. Philos. Logic* **1**, 61–69.

Hannequin, A. (1895), *Essai Critique sur l'Hypothèse des Atomes dans la Science Contemporaine*, Masson, Paris.

Jourdain, P.E.B. (1912), Mr. Bertrand Russell's first work on the principles of mathematics, *Monist* **22**, 149–158.

Jourdain, P.E.B. (1913), A correction and some remarks, *Monist* **23**, 145–148.

Jourdain, P.E.B. (trans.) (1915), *Cantor, Contributions to the Founding of the Theory of Transfinite Numbers, with an Introduction and Notes*, Open Court, New York. (Reprinted by Dover, New York, 1955.) (English translation of Cantor (1895).)

Kleene, S.C. (1967), *Mathematical Logic*, Wiley, New York.

Kleene, S.C. (1971), *Introduction to Metamathematics*, North-Holland/American Elsevier, Amsterdam/New York.

Kneale, W. and Kneale, M. (1962), *The Development of Logic*, Clarendon Press, Oxford.

Levy, A. (1979), *Basic Set Theory*, Springer-Verlag, New York.

Moore, G.H. (1982), *Zermelo's Axiom of Choice: Its Origins, Development and Influence*, Springer-Verlag, New York.

Moore, G.H. (1988), The roots of Russell's paradox, *Russell* (n.s.) **8**, 46–56.

Moore, G.H. and Garciadiego, A.R. (1981), Burali-Forti's paradox: a reappraisal of its origins, *Historia Math.* **8**, 319–350.

Nidditch, P.H. (1962), *The Development of Mathematical Logic*, Routledge & Kegan Paul, London; Dover, New York.

Peano, G. (1902–1906), Super teorema de Cantor–Bernstein et additione, *Riv. Mat.* **8**, 136–157.

Poincaré, H. (1906), Les Mathématiques et la logique. IV. *Rev. Métaphysique et de Morale* **14**, 294–317.

Rosser, J.B. (1953), *Logic for Mathematicians*, McGraw-Hill, New York.

Russell, B. (1983), *What Shall I Read?* MS notebook, AR. Covers the years 1891–1902. (In K. Blackwell, N. Griffin, R. Rempel, and J. Slater (eds.), *Collected Papers of Bertrand Russell, I*, Allen and Unwin, London, pp. 347–365.)

Russell, B. (1897), *Review of De l'Infini Mathématique* by Louis Couturat, AR-BREP 2:6. *Mind* **6**, 112–119.

Russell, B. *Russell–Couturat Correspondence*, 1897–1914, AR.

Russell, B. (1901), *Recent work on the Principles of Mathematics*, AR-BREP 2:39, 15 pp. Published in *International Monthly* **4**, 83–101. (Reprinted as *Mathematics and the Metaphysicians*, B. Russell (1918), *Mysticism and Logic*, Longmans, Green, New York, pp. 74–96.)

Russell, B. (1902), *Letter to Frege (June 16, 1902)*, in German; AR. (English translation by B. Woodward in van Heijenoort (1967a, pp. 124–125).)

Russell, B. (1903), *Principles of Mathematics*, Cambridge University Press, London.

Russell, B. (1919), *Introduction to Mathematical Philosophy*, Allen and Unwin, London.

Temple G. (1981), *100 Years of Mathematics*, Springer-Verlag, New York.

van Heijenoort, J. (ed.) (1967a), *From Frege to Gödel: A Source Book in Mathematical Logic, 1879–1931*. Harvard University Press, Cambridge, MA.

van Heijenoort, J. (1967b), Logical paradoxes, in P. Edwards (ed.), *Encyclopedia of Philosophy*, Vol. 5, Macmillan, New York, pp. 45–51.

Zermelo, E. (ed.) (1932), *Cantors Gesammelte Abhandlungen mathematischen und philosophischen Inhalts*, Springer-Verlag, Berlin. (Reprinted: Olms, Hildesheim, 1962.)

Principia Mathematica and the Development of Automated Theorem Proving

Daniel J. O'Leary

1. Introduction

The present chapter describes the first two major published works in automated theorem proving, the *Logic Theory Machine* (LT) of Newell, Shaw, and Simon about 1956 and the work of Wang in 1958. Both works attempted to prove theorems in propositional logic found in *Principia Mathematica*. While the LT is the first published example of automated theorem proving, Martin Davis, in 1954, wrote a program to prove theorems in additive arithmetic (Loveland, 1984, p. 1). His results were not published.

Part I, Section A of *Principia Mathematica* (PM) (Whitehead and Russell, 1935) is entitled "The Theory of Deduction" and covers 35 pages of the sixteen-hundred page work. In this section, Whitehead and Russell state and prove about two hundred theorems in propositional logic. The LT attempted to prove the first 52 theorems and succeeded with 38. Wang's program proved them all. The two projects had the same goal, but quite different approaches. Even the concept of what constitutes a proof is different.

To help understand the differences in these approaches we adopt some definitions from Loveland (1984). By *automated theorem proving* we mean "the use of a computer to prove nonnumerical results ... often [requiring] human readable proofs rather than a simple statement 'proved'." Loveland identifies two major approaches to automated theorem proving which he calls the *logic approach* and the *human simulation* or *human-orientated* approach. "The logic approach is characterized by the presence of a dominant logical system that is carefully delineated and essentially static over the development stage of the theorem proving system."

The human simulation programs attempt to find proofs in the way a mathematician would. They usually have a body of facts, a knowledge base to which they refer, an executive director which controls the decision-making process, and the ability to chain both forward and backward. Newell, Shaw, and Simon's LT is a human simulation theorem prover; it attempts to find a proof a human would find using all the theorems proved previously. Wang's work follows the logic approach, and tends to give a result more akin to the word "proved" with reasons why the assertion can be made.

2. The Logic Theory Machine (LT)

Herbert Simon and Allen Newell of Carnegie-Mellon University and J.C. Shaw of the Rand Corporation wrote the LT programs in 1956 and 1957. They wrote this human simulation program to understand "how it is possible to solve difficult problems such as proving mathematical theorems, discovering scientific laws from data, playing chess, or understanding the meaning of English prose." (Newell et al., 1956, p. 109) They state:

> The research reported ... is aimed at understanding the complex processes (heuristics) that are effective in problem solving. Hence, we are not interested in methods that guarantee solutions, but which require vast amounts of computation. Rather, we wish to understand how a mathematician, for example, is able to prove a theorem even though he does not know when he starts how, or if, he is going to succeed.

2.1. Inside the Logic Theory Machine

The LT has three methods available to prove theorems: substitution, detachment, and chaining. The user gives the LT a theorem to prove and all theorems in PM up to that point. The LT may not have proved any of them. The theorems which precede the one to be proved are placed on a theorem list. The program refers to this theorem list as it attempts to apply various proof techniques.

The substitution method attempts to transform an entry on the theorem list to the desired theorem. It may substitute variables or replace connectives based upon definitions. The detachment method makes use of *modus ponens*. Assume we wish to prove B. The LT searches the theorem list for an entry of the form A implies B. If one is found A becomes a subproblem which the LT adds to a subproblem list. Chaining falls into two types, forward and backward. Both methods rely on the transitivity of implication. Assume the LT wishes to prove A implies C. In forward chaining the LT searches the theorem list for an entry of the form A implies B and if found, enters B implies C on the subproblem list. In backward chaining the LT searches the theorem list for B implies C and enters A implies B on the subproblem list.

The LT contains an executive director which controls the use of the three methods. The executive director uses the following scheme:

First use substitution.

Should substitution fail, try detachment. The result of each detachment is submitted to the substitution method. If substitution fails, then the result of detachment is placed on the subproblem list.

Try forward chaining next and submit the result to substitution. If a proof has not been found, the result of forward chaining is added to the subproblem list.

Try backward chaining in a manner similar to forward chaining.

When all the methods have been tried on the original problem, the executive director goes to the subproblem list and chooses the next subproblem. The techniques are applied in the same sequence, and if no proof is found the next problem on the subproblem list is tried.

The search for a proof stops when one of four conditions have been met:

1. the LT has found a proof;
2. the bounds set on time for the search have been exceeded;
3. the bounds set on memory for the search have been exceeded; or
4. no untried problems remain on the subproblem list.

2.2. An Example of a Logic Theory Proof

One of the theorems LT proved is

$$*2.45 \quad \vdash: \sim(p \lor q). \supset . \sim p.$$

It is instructive to compare the proofs of LT, PM, and Wang's program. In PM the proofs are often abbreviated. The authors (Whitehead and Russell, 1935, p. vi), state "The proofs of the earliest propositions are given without omission of any step, but as the work proceeds the proofs are gradually compressed, retaining however sufficient detail to enable the reader by the help of the references to reconstruct proofs in which no step is omitted."

The proofs of *2.45 in PM is presented as:

$$*2.45 \quad \vdash: \sim(p \lor q). \supset . \sim p \qquad [2.2, \text{Transp}].$$

Transp refers to one of seven theorems which carry that name. The proof given by LT and the reconstruction from PM are identical. (Wang's proof is given below.) The reconstruction of the proof is:

*2.45	$\sim(p \lor q). \supset . \sim p$	
1.	$p. \supset . p \lor q$	*2.2
2.	$p \supset q. \supset . \sim q \supset \sim p$	*2.16 [Transp]
	$q/p \lor q$	
3.	$p \supset . p \lor q: \supset : \sim(p \lor q). \supset . \sim p$	Sub. in 2
4.	$\sim(p \lor q). \supset . \sim p$	Modus ponens 3, 1

By examining the proof, one may trace the successful path taken by LT. The theorem list contains all theorems up to *2.45. Since the consequent of *2.16 has the same structure as the theorem to be proved, a substitution will make them identical (see lines 2 and 3 in the proof). The LT then tries to find a match for the antecedant and locates *2.2 on the theorem list.

3. Wang's Techniques

Newell, Shaw, and Simon contrasted algorithmic and heuristic methods in searching for a proof. The algorithmic methods guarantee that if a problem has a solution it will be found. No claim, however, is made on how long it will take or the resources required. In the heuristic method we give up the guarantee, but solutions may be found at reduced cost.

To illustrate the algorithmic method, Newell, Shaw, and Simon put forward the British Museum Algorithm—look everywhere. All axioms are proofs of length 1. All proofs of length n are formed by making all the permissible substitutions and replacements in proofs of length $(n - 1)$ and all permissible detachments of pairs of theorems generated up to this point. This method will, of course, generate every theorem of PM, but at a high cost.

Wang thought the argument against the algorithmic method, taking the British Museum Algorithm as an example, was weak. He wanted to show that there were other algorithmic methods available and the British Museum Algorithm only a straw man. Wang sought a method that guaranteed a proof and did not require extensive running time.

3.1. The Wang Program

Wang's algorithm, based on sequent logic, is much better than the British Museum Algorithm. Five connectives are used in PM: negation, implication, disjunction, conjunction, and equivalence. Wang's system uses eleven rules of inference. One rule identifies a theorem, five introduce a connective on the left-hand side of the sequent arrow, and five introduce a connective on the right-hand. These rules are stated below:

P1. Initial rule: If λ, ζ are strings of atomic formulas, then $\lambda \rightarrow \zeta$ is a theorem if some atomic formula occurs on both sides of the arrow.

In the following ten rules, λ and ζ are always strings (possibly empty) of atomic formulas.

P2a. Rule $\rightarrow \sim$: If $\phi, \zeta \rightarrow \lambda, \rho$, then $\zeta \rightarrow \lambda, \sim \phi, \rho$.
P2b. Rule $\sim \rightarrow$: If $\lambda, \rho \rightarrow \pi, \phi$, then $\lambda, \sim \phi, \rho \rightarrow \pi$.
P3a. Rule $\rightarrow \&$: If $\zeta \rightarrow \lambda, \phi, r$ and $\zeta \rightarrow \lambda, \psi, \rho$, then $\zeta \rightarrow \lambda, \phi \& \psi, \rho$.
P3b. Rule $\& \rightarrow$: If $\lambda, \phi, \psi, \rho \rightarrow \pi$, then $\lambda, \phi \& \psi, \rho \rightarrow \pi$.
P4a. Rule $\rightarrow \vee$: If $\zeta \rightarrow \lambda, \phi, \psi, \rho$, then $\zeta \rightarrow \lambda, \phi \vee \psi, \rho$.
P4b. Rule $\vee \rightarrow$: If $\lambda, \phi, \rho \rightarrow \pi$, then $\lambda, \phi \vee \psi, \rho \rightarrow \pi$.
P5a. Rule $\rightarrow \supset$: If $\zeta, \phi \rightarrow \lambda, \phi, \rho$, then $\lambda \rightarrow \phi \supset \psi, \rho$.
P5b. Rule $\supset \rightarrow$: If $\lambda, \psi, \rho \rightarrow \pi$ and $\lambda, \rho \rightarrow \pi, \phi$, then $\lambda, \phi \supset \psi, \rho \rightarrow \pi$.
P6a. Rule $\rightarrow \equiv$: If $\phi, \zeta \rightarrow \lambda, \psi, \rho$ and $\psi, \zeta \rightarrow \lambda, \phi, \rho$, then $\zeta \rightarrow \lambda, \phi \equiv \psi, \rho$.
P6b. Rule $\equiv \rightarrow$: If $\phi, \psi, \lambda, \rho \rightarrow \pi$ and $\lambda, \phi \rightarrow \pi, \phi, \psi$, then $\lambda, \phi \equiv \psi, \rho \rightarrow \pi$.

In sequent logic a theorem is proved by taking a finite set of strings in the form specified by rule P1 and applying the other rules until the theorem is proved. Wang's program turns this process around. The applications of the inference rules were deduced by starting at the theorem until a finite number of occurrences of rule P1 are reached. Given a theorem in PM to prove, one places a sequent arrow in front of it and proceeds. If, however, the main connective is an implication, Wang often replaces it with the sequent arrow and then starts.

3.2. Examples Of Proofs

The proof of *2.45 generated by this method is shown below:

$$*2.45 \quad {\sim}(p \vee q) \to {\sim}p \qquad\qquad\qquad (1)$$

$$(1) \quad \to {\sim}p, p \vee q \qquad\qquad\qquad\qquad (2)$$

$$(2) \quad p \to p \vee q \qquad\qquad\qquad\qquad (3)$$

$$(3) \quad p \to p, q \qquad\qquad\qquad\qquad\quad (4)$$

The numbers on the right-hand side are line numbers. The numbers on the left-hand side show which line was used to produce the line of interest. The connectives are dealt with in order, from left to right. Negation, the left-most connective of line (1) can be introduced on the left-hand side of a sequent arrow by rule P2b, so line (1) could be generated by something of the form in line (2). The lines are examined until the conditions of rule P1 are satisfied.

Another example is given by

$$*5.21 \quad \vdash: {\sim}p \,\&\, {\sim}q \,.\, \supset \,.\, p \equiv q$$

a proof of which is shown below.

$$*5.21 \quad \to {\sim}p \,\&\, {\sim}q \,.\, \supset \,.\, p \equiv q \qquad\qquad (1)$$

$$(1) \quad {\sim}p \,\&\, {\sim}q \to p \equiv q \qquad\qquad\qquad (2)$$

$$(2) \quad {\sim}p, {\sim}q \to p \equiv q \qquad\qquad\qquad (3)$$

$$(3) \quad {\sim}q \to p \equiv q, p \qquad\qquad\qquad (4)$$

$$(4) \quad \to p \equiv q, p, q \qquad\qquad\qquad\quad (5)$$

$$(5) \quad p \to q, p, q \qquad\qquad\qquad\qquad (6)$$

$$(5) \quad q \to p, p, q \qquad\qquad\qquad\qquad (7)$$

The connectives are dealt with from left to right, across the proof line as before. Finally, at line (5) there is an equivalence on the right-hand side of the arrow. By Rule P6b this must have come from two formulas. One is found in line (6) and the other in line (7).

The proof is always in the form of a finite tree, and the program conducts a depth-first search through the tree. A branch is explored until all connectives are eliminated. The program then checks that rule P1 is satisfied. The program stops when all branches have been explored. The formula submitted is a theorem if rule P1 is satisfied at the bottom of each branch of the search tree. The program does not produce proofs in the sense of PM, since it does not refer to previously proved theorems, nor is there a progression of theorems from a set of axioms.

4. Sidelights of the Logic Theory Machine

Simon (1956) wrote to Russell in late 1956 and described the work of the LT. Russell (1956) replied, "I am delighted to know that *Principia Mathematica* can now be done by machinery. I wish Whitehead and I had known of this possibility before we wasted ten years doing it by hand. I am quite willing to believe that everything in deductive logic can be done by machinery." The LT was named as a co-author on a paper submitted to the *Journal of Symbolic Logic*, but the paper was refused (Tiernay, 1983).

The proof of *2.85 given in PM, shown below, contains an error. The LT was able to find a remarkable proof of *2.85.

*2.85 $\vdash: . p \vee q . \supset . p \vee r : \supset : p . \vee . q \supset r$

Dem.

[Add . Syll] $\vdash: . p \vee q . \supset . r : \supset . q \supset r$ (1)

$\vdash . *2.55 . \supset \vdash:: {\sim}p . \supset : . p \supset r . \supset . r : .$

[Syll] $\supset : . p \vee q . \supset . p \vee r : \supset : p \vee q . \supset . r : .$

[(1) . *2.83] $\supset : . p \vee q . \supset . p \vee r : \supset : q \supset r$ (2)

$\vdash . (2) . \text{Comm} . \supset \vdash: . \quad p \vee q . \supset . p \vee r : \supset : {\sim}p . \supset . q \supset r :$

[*2.45] $\supset : p . \vee . q \supset r : . \supset \vdash . \text{Prop}$

The line numbered (2) cannot be obtained as indicated. The intent of Whitehead and Russell here is not clear, as explained in an unpublished paper from the Russell Conference of 1984. Simon (1956) wrote to Russell stating that the LT had "created a beautifully simple proof to replace a far more complex one in [*Principia Mathematica*]." The proof sent to Russell, however, contains a simple error. The correct proof is shown below:

*2.85 $p \vee q . \supset . p \vee r : \supset : p . \vee . q \supset r$
1. $p \supset q . \supset : q \supset r . \supset . p \supset r$ *2.06
   $p/q, q/p \vee q, r/p \vee r$
2. $q \supset p \vee q . \supset : p \vee q \supset p \vee r . \supset . q \supset p \vee r$ Sub. 1
3. $q \supset p \vee q$ *1.3
4. $p \vee q \supset p \vee r . \supset . q \supset p \vee r$ *Modus ponens* 3, 2
5. $p \vee q \supset p \vee r . \supset . {\sim}q \vee (p \vee r)$ Def. of $\supset$
6. $p \vee (q \vee r) . \supset . q \vee (p \vee r)$ *1.5
   $p/{\sim}q, q/p$
7. ${\sim}q \vee (p \vee r) . \supset . p \vee ({\sim}q \vee r)$ Sub. 6
8. $p \vee q \supset p \vee r . \supset . p \vee ({\sim}q \vee r)$ Chain with 5 and 7
9. $p \vee q \supset p \vee r . \supset . p . \vee . q \supset r$ Def of $\supset$

In his letter Simon states, "The machine's proof is both straightforward and unobvious, we were much struck by its virtuosity in this instance."

5. Conclusion

Principia Mathematica is clearly a human creation. Its formalism, however, almost begs for a mechanical procedure. There is no surprise that, when looking for some human endeavor to simulate with a computer, PM is chosen. Both methods described here have had a profound effect on automated theorem proving. The work of Newell, Shaw, and Simon laid a strong foundation for the modern techniques of artificial intelligence. The work of Wang immediately opened up a new front in the battle to understand what computers can do.

Bibliography

Loveland, D.W. (1984), Automated theorem proving: a quarter-century review, in *Automated Theorem Proving* (Bledsoe and Loveland, eds.), American Mathematical Society, Providence, RI, pp. 1–45.

Newell, A., Shaw, J.C., and Simon, H.A. (1956), Empirical explorations of the logic theory machine: A case study in heuristics. *Proceedings of the Western Joint Computer Congress*, 1956, pp. 218–239. Also in *Computers in Thought* (Fiegenbaum and Feldman, eds.), McGraw-Hill New York, 1963, pp. 134–152. Page numbers cited are from this reference.

O'Leary, D.J. (1984), The Propositional Logic of *Principia Mathematica*, and Some of its Forerunners. Presented at the Russell Conference, 1984, Trinity College, University of Toronto. Unpublished.

Russell, B. (1956), Letter to Herbert Simon dated 2 November 1956. Located at the Russell Archives, McMaster University.

Simon, H. (1956) Letter to Bertrand Russell dated 9 September 1957. Located at the Russell Archives, McMaster University.

Tiernay, P. (1983), Herbert Simon's simple economics. *Science* 83, **4**, No. 9, November 1983.

Wang, H. (1960), Toward mechanical mathematics, *IBM J. Res. Develop.*, 1960, 2–22. Also in *Logic, Computers and Sets*, Chelsea, New York, 1970.

Whitehead, A.N. and Russell, B. (1935), *Principia Mathematica*, 2nd ed., Cambridge University Press, Cambridge, UK.

Oswald Veblen and the Origins of Mathematical Logic at Princeton*

William Aspray

The remarkable transformation from Aristotelean to mathematical logic in the period 1880–1930 was largely a European affair. The important centers of change were all located in Europe.[1] Few Americans pursued research in logic, and those who did seldom contributed to mainstream developments.[2] By the end of the 1930s, however, the United States had developed an important, indigenous research program in mathematical logic.[3] For the first time American faculties were training their own students, making their own research contributions, and publishing their own research journals at levels competitive with the best Europe had to offer.

In the 1930s Princeton became a leading center for mathematical logic, both nationally and internationally. Its role in the formation of recursive function theory is well known, as are the contributions of its leading logicians, Alonzo Church and Kurt Gödel. The combined mathematical faculties of Princeton University and the Institute for Advanced Study formed a potent research nucleus that attracted a regular flow of national and international visitors, including many interested in logic: Paul Bernays, Haskell Curry, Kurt Gödel, Karl Menger, Emil Post, Alan Turing, and Stanislaw Ulam. The university bountifully produced strong Ph.D.s in mathematical logic: Alonzo Church, A.L. Foster, J. Barkley Rosser, Stephen Kleene, Alan Turing, Leon Henkin, and John Kemeny—a number of whom became leaders of mathematical logic in the post-war generation.[4] At the same time, Princeton raised the *Journal of Symbolic Logic* to its present level of international distinction.

Others have written about the development of recursive function theory and the contributions of Church and Gödel.[5] This chapter discusses instead the preparations for this golden era in Princeton. It is primarily the story of Oswald Veblen, his use of mathematical logic to pursue a secure foundation

* I appreciate the critical reading of early drafts of this paper by Thomas Drucker, Sheldon Hochheiser, Gregory Moore, and Albert Tucker; information provided by Irving Anellis, John Corcoran, and Dale Johnson; and research assistance from Robbin Clamons and Marek Rostocki.

for geometry, and his efforts to establish an institutional context in Princeton conducive to mathematical research.

1. From Geometry to Postulate Theory

Veblen did not contribute to the discipline of mathematical logic as we know it today, but instead to a transitional discipline known generally as American Postulate Theory. His work, together with that of E.V. Huntington, E.H. Moore, Edwin Wilson, and several others at the turn of this century, investigated the logical adequacy of axiom systems for mathematics.[6] For the Postulate Theorists, logic was a subject worth developing only so far as it contributed to the rigorous, i.e., axiomatic, advancement of working mathematics. The group developed concepts of completeness, independence, and categoricity in order to apply them to axiom systems for algebra, geometry, and analysis.

Veblen was first and foremost a geometer. Indeed, he was the leading American geometer of the period, with fundamental contributions to projective geometry, topology (*analysis situs*), differential geometry, and spinors. His interest in logic followed naturally out of an interest in the foundations of geometry, so it is not surprising that his contributions to logic are reported mainly in his publications on geometry.

After completing undergraduate degrees at Iowa and Harvard in 1910, Veblen entered the graduate mathematics program at the University of Chicago. There he came under the influence of Oskar Bolza, Heinrich Maschke, and particularly E.H. Moore, who directed his dissertation research. Moore had taken a strong interest in the foundations of geometry at about the time Veblen arrived at Chicago, because of the publication in 1899 of David Hilbert's *Grundlagen der Geometrie* and the recent foundational research of Moritz Pasch and Giuseppe Peano. In 1901 Moore wrote his most significant paper on the foundations of geometry,[7] and in the 1901–1902 academic year presented a course and research seminar on the subject, which Veblen attended.[8]

Veblen's first professional publication was a comparison of Hilbert's logical foundations in *Grundlagen der Geometrie* with those provided in lesser-known works by Levi-Civita, Pasch, Peano, and Veronese. Veblen points out that, despite the numerous enthusiastically positive reviews of Hilbert's work, which sought to correct the logical inadequacies of Euclid's *Elements*, the *Grundlagen* itself left room for logical improvement. In particular, Veblen shows that the number of Hilbert's undefined terms can be reduced, that Hilbert makes a logical error in demonstrating the independence of his axioms of congruence, and (reiterating a result of E.H. Moore) that one of Hilbert's axioms can be deduced from the other axioms.

Veblen continued these investigations in his doctoral dissertation, which he completed in 1903 and published in 1904.[9] The dissertation presents an axiomatic system for three-dimensional Euclidean geometry. His system follows the approach of Pasch and Peano, rather than that of Hilbert and Pieri.[10] It is based on two undefined terms, "point" and "order," and only twelve axioms.[11] Veblen devotes considerable attention to the logical adequacy of his system. He carefully distinguishes in the modern way between axioms and derived statements, and between axioms and definitions.[12] After a general discussion of the importance of independence in axiom systems, he gives examples that establish the independence of each of his axioms. He also shows how each class of objects satisfying his axioms can be coordinatized, and how two such classes can be placed in one-to-one correspondence by identifying their coordinate systems, so as to establish the categoricity of the axiom system.[13]

Veblen remained at Chicago as an associate for two years after his degree was granted, during which time he helped to advise R.L. Moore's dissertation on the foundations of Euclidean geometry. After his move to Princeton in 1905, Veblen focused first on foundational issues in real analysis and projective geometry.[14] One paper presented a categorical and independent system of axioms for order properties in the linear continuum and in well-ordered sets.[15] In analysis, Veblen presented the first rigorous proof of the Jordan Curve Theorem and continued his study of the nature of compactness.[16] The Heine–Borel Theorem, which was one of the axioms presented in his dissertation, became the basis for the first "rigorous and systematic treatment in English on theory of functions of a real variable."[17]

The research on the foundations of projective geometry were carried out by Veblen in collaboration with his Princeton colleague, John Wesley Young. In 1908 they presented the first independent set of axioms for projective geometry, for spaces with real or complex coordinates.[18] Their work culminated in the two-volume treatise *Projective Geometry*, which served as the standard advanced textbook for many years.[19] By adopting an algebraic approach, in which they considered the group of transformations associated with the geometry, they were able to extend Klein's Erlanger program to projective geometry.[20] They were attentive to the logical properties of their axiom system and attempted "not merely to prove every theorem rigorously but to prove it in such a fashion as to show in which spaces it is true and to which geometries it belongs."[21]

Although their primary purpose in these volumes was pedagogical, they devoted considerable attention to foundational issues. The first chapter of volume I discusses the nature of undefined elements, unproved propositions, consistency, categoricity, independence, and the logical properties of a formal mathematical system. They defended this approach in the Preface to the first volume in a way that revealed Veblen's hope for a unified treatment of and foundation for all of geometry:

> Even the limited space devoted in this volume to the foundations may seem a drawback from the pedagogical point of view of some mathematicians. To this we can only reply that, in our opinion, an adequate knowledge of geometry cannot be obtained without attention to the foundations. We believe, moreover, that the

abstract treatment is peculiarly desirable in projective geometry, because it is through the latter that the other geometric disciplines are most readily coordinated. Since it is more natural to derive the geometrical disciplines associated with the names of Euclid, Descartes, Lobatchewsky, etc., from projective geometry than it is to derive projective geometry from one of them, it is natural to take the foundations of projective geometry as the foundations of all geometry.[22]

Veblen's later research, in *analysis situs* (topology), differential geometry, and spinors, showed the same interest in foundational issues as his earlier work.[23] A paper of 1913 co-authored with James Alexander set forth a logically organized and rigorous account of Henri Poincaré's memoirs on *analysis situs*. A paper of 1931 co-authored with J.H.C. Whitehead presented a set of axioms for differential geometry of *n*-dimensional manifolds.[24] Veblen struggled for several decades to extend Klein's Erlanger program to all of geometry, including topology and differential geometry. He was never successful in unifying the foundation for geometry, and in the end had to admit that a branch of mathematics should be called *geometry* "because the name seems good, on emotional and traditional grounds, to a sufficient number of competent people."[25]

One last study of foundations by Veblen deserves mention. In his 1924 presidential address to the American Mathematical Society, Veblen placed the foundations of geometry into the larger context of the foundations of mathematics.[26] He notes that, although geometry and real analysis were being taught from an arithmetical point of view, "the arithmetical definition of mathematics ... hardly seems adequate when we examine its foundations ... the whole arithmetical structure is a portion of the edifice of formal logic which we are not as yet able to separate off."[27] He points out the contributions of Russell and Whitehead to the logic of propositions and the logic of classes, and suggests that their work leads to the arithmetization of mathematics (which he attributes to Kronecker and Study). He notes the problem this arithmetization program has with the definition of irrational number, the theory of infinite classes, and the set-theoretic paradoxes, and points to Hilbert's program as "the latest and, I think, most promising" attempt to resolve these difficulties. He closes the address by stating:

> Whether or not this is a correct statement of Hilbert's program, the conclusion seems inescapable that formal logic has to be taken over by mathematicians. The fact is that there does not exist an adequate logic at the present time, and unless the mathematicians create one, no one else is likely to do so. In the process of constructing it we are likely to adopt the Russell view that mathematics is coextensive with formal logic. This, I suppose, is apt to happen whether we adopt the formalist or the intuitionalist point of view.[28]

2. Veblen as Institution Builder

The most enduring names in mathematics have generally been those of the great researchers. Veblen's research contributions to logic do not match those of Peano, Zermelo, or Gödel; and, on these criteria, it seems just that Veblen's name has been all but lost from the history of logic. But we must remember that to flourish even mathematics, generally regarded as the scientific discipline least dependent on its social context, has certain requirements: financial support and research time for its practitioners, good educational programs and libraries, a community of knowledgeable practitioners, and communication channels such as professional meetings and journals. In this light, Veblen's contributions appear more significant, for in addition to being an able researcher he was one of the chief builders of strong American mathematical research institutions. These efforts resulted in the growth of American logic, particularly in Princeton.

The story begins with Veblen's appointment in 1905 as a preceptor in the Princeton University Mathematics Department.[29] Only 10 years earlier the institution had been known as The College of New Jersey and was an intellectual backwater, with a nonresearch mathematics faculty offering a curriculum exclusively of elementary courses. Within that decade the college adopted its present name, established graduate education, and hired Woodrow Wilson as president to reform the educational program. Wilson established a preceptorial program at Princeton, which resulted in smaller classes and more personalized instruction—a program that was strongly supported by Henry Burchard Fine, chairman of the mathematics department and dean of the faculty. Fine used the extra faculty positions the preceptorial system required to hire promising young research mathematicians without regard to nationality—a hiring practice not followed by other American universities. Veblen, together with Gilbert Bliss, Luther Eisenhart, and John Wesley Young were hired in 1905, the first year of the preceptorial program. Over the next 20 years, Fine used these positions to hire other promising young research mathematicians: James Alexander, G.D. Birkhoff, Pierre Boutroux, Thomas Gronwall, Einar Hille, Solomon Lefschetz, and J.H.M. Wedderburn.

Veblen's career advanced rapidly. He was promoted to full professor in 1910 and became one of three organizational and intellectual leaders of the department, together with Fine and Eisenhart. His books on *Projective Geometry* (1910; 1918) and *Analysis Situs* (1922) were recognized as the leading works in their fields. Through the 1910s and early 1920s he held increasingly important positions in the professional communities of American mathematics and science: member, American Philosophical Society, 1912; vice-president, American Mathematical Society (AMS), 1915; AMS Colloquium Lecturer, 1916; vice-president, Mathematics Association of America, 1917; member, National Academy of Science, 1919; various senior positions, National Research Council (NRC), Division of Physical Sciences, 1920–23; vice-president, American Academy of Arts and Sciences, 1921; and president, AMS, 1923–24.

Veblen was an activist reformer during his term as AMS president. He established an endowment fund for the society and raised money to subsidize a stronger publication program. Using to advantage his positions within the

AMS and NRC, he arranged for mathematicians to be supported under the NRC fellowship program, which had been limited previously to physicists and chemists. These actions enabled promising American research mathematicians to hold post-doctoral positions at leading centers of mathematical research in Europe and America, and to publish their results in American journals.

Fine, Veblen, and other Princeton mathematicians continued to pursue their independent research careers as time permitted, but through the early 1920s Princeton remained primarily an undergraduate teaching institution. Heavy undergraduate teaching loads prevailed, most faculty had to work at home for lack of offices, and little money was available to improve facilities or research opportunities.

Veblen sought to rectify this situation with a proposal for an Institute for Mathematical Research, which he submitted between 1924 and 1926 to the National Research Council and the General Education Board of the Rockefeller Foundation. His proposal called for an institute of four or five senior mathematicians and an equal number of junior colleagues, who would devote themselves "entirely to research and to the guidance of the research of younger men."[30] These mathematicians were to be released from all usual teaching and committee duties, though they were to "be free to offer occasional courses for advanced students."[31] Believing it to be a further incentive to funding, Veblen proposed the institute be dedicated to applied mathematics—a subject he regarded as under-represented in the United States.

Neither organization supported Veblen's plan, but it influenced the Princeton University fund drive for the sciences that Dean Fine organized in 1926 for President Hibben. In a fund-raising document Fine outlined the needs of the mathematics department, many of them expressed earlier by Veblen: endowed research professorships, increased staff with lighter teaching loads allowing time for research, offices and other facilities for the mathematical community, graduate scholarships, continued support for the journal *Annals of Mathematics* edited at Princeton, and a research fund for use to meet changing conditions. The Rockefeller Foundation awarded Princeton one million dollars for this research fund, which the university matched by 1928 with two million dollars in alumni gifts. One-fifth of the total, six hundred thousand dollars, was allocated to the mathematics department.

These alumni gifts included chairs in mathematics and mathematical physics endowed by an old friend of Fine, Thomas Jones, and Jones's brother David and niece Gwenthalyn. Veblen held the chair in mathematics from 1926 until his resignation from the university in 1933.

In 1928, Jones and his niece provided additional funds to build and maintain a mathematics building. Veblen assumed responsibility for the design of a building, as Jones said, "any mathematician would be loath to leave."[32] Fine Hall housed private faculty studies (many generously appointed with carved oak paneling, fireplaces, leather sofas, and oriental rugs), a well-designed and well-stocked mathematics and physics library, three classrooms designed to the specific needs of mathematics instruction, and two common rooms (with kitchens) for social and intellectual interaction of the faculty, graduate students, and visitors.[33] Veblen carefully planned each feature to promote research and the interaction he believed was important to sustain it. In his own words, he wanted a center "about which people of like intellectual interests can group themselves for mutual encouragement and support, and where the young recruit and the old campaigner can have those informal and easy contacts that are so important to each of them."[34] The design was tremendously successful; Fine Hall became a place that attracted a large community of graduate students, faculty, post-doctoral fellows, and visiting mathematicians to interact on both social and intellectual levels—a "grand hotel of mathematics" as one faculty member called it.[35]

By the time the mathematics department moved into Fine Hall in 1931, all of the needs of the mathematics department outlined in Fine's 1926 fund-raising document had been fulfilled, except for a larger professional staff with time for research. This need was met, and the overall situation improved further, with the founding of the Institute for Advanced Study. The story of its founding is told elsewhere, and need not be told here.[36] What is less well known is Veblen's role in shaping the institute and, in doing so, improving the overall strength of mathematics at Princeton.

Veblen was hired in 1932 as the first faculty member of the institute. He was the chief architect of the institute's research environment, and served informally as a behind-the-scenes administrator—"the power behind the throne" as one admiring colleague referred to him.[37] Even after his retirement, in 1950, he continued to exert a profound influence on the affairs of the institute, as a member of its board of trustees.

He was largely responsible for hiring the other original members of the faculty: James Alexander, Albert Einstein, Marston Morse, John von Neumann, and Hermann Weyl. The departure of Alexander, Veblen, and von Neumann for the institute, and of Einar Hille for Yale, seriously threatened the university's research strength in mathematics that had been carefully nurtured over the previous quarter century. Sensitive to the potential damage to the university and the institute's need for a good research library, physical facilities, and a critical mass of trained mathematicians, Veblen and Eisenhart arranged for the institute to share quarters with the department in Fine Hall—an arrangement which continued until the institute built Fuld Hall a mile from the university campus in 1939.[38]

In the early years the direction of the institute was uncertain. Its charter established it as an educational organization with the right to grant doctoral degrees. Some instead envisioned the institute as offering a protected research environment, where advanced researchers could pursue their work without teaching duties or other distractions. Under Veblen's direction, the institute steered a third course which emphasized both post-doctoral education and advanced research. In effect, the Institute for Advanced Study realized Veblen's earlier plan for an Institute for Mathematical Research. The senior faculties of the institute were not encumbered with committee or teaching duties, though they sometimes elected to teach graduate or post-graduate seminars open to both the institute and the university communities. There was ample support to bring visitors to the institute to pursue their research. The institute was a particular attraction to young mathematicians of research promise, a number of whom served as research assistants to the permanent faculty.[39] Especially in the 1930s, when the institute and the department were housed together in Fine Hall, there were ample opportunities for graduate students to interact with the institute faculty and visitors; and a number of dissertations were informally directed by institute personnel.[40]

In the end, the founding of the institute greatly strengthened mathematics in Princeton. Many new mathematical appointments were made, including those to Kurt Gödel at the institute and Alonzo Church at the university. In the 1930s the university awarded 39 Ph.D.s in mathematics, among which were those to A.L. Foster, Stephen Kleene, J. Barkley Rosser, Alan Turing, and J.H.C. Whitehead (all of whom studied with Church or Veblen). Several hundred distinguished mathematicians from around the world were long-term visitors to Princeton in this decade. With the careful shepherding of Veblen and others, Princeton had become what the Danish mathematician Harald Bohr called "the mathematical center of the world."[41]

3. Logic at Princeton

Veblen was interested in logic as a branch of mathematics, and he believed that general strength in mathematics contributed to strength in logic. This belief was confirmed by what happened in Princeton. The strong faculty, graduate program, and fellowship support attracted a strong pool of graduate students. In a number of cases the students who wrote dissertations on problems in logic chose Princeton for graduate study not because of any special interest in logic, but because of the overall quality of its mathematics program.[42] This same quality attracted the many distinguished visitors from the United States and abroad. The impact this had on logic is aptly aphorized by Rosser's declaration that four-fifths of American logic was at Princeton.[43]

Veblen adamantly held that Princeton's in-house editing of professional journals contributed markedly to the mathematics program; that the journals provided the local community with an outlet for publication, a wide set of contacts in the greater mathematical community, early information on new mathematical talent, and some control over the direction of research (especially American research). So Veblen pushed Princeton to provide funds to support *Annals of Mathematics*, *Annals of Mathematics Studies*, *Journal of Symbolic Logic*, and *Annals of Mathematical Statistics*, and to house their editorial offices.

Although the Association for Symbolic Logic and its organ, the *Journal of Symbolic Logic*, had their origins in the Philosophy Department at Brown

University, Princeton contributed to their early development.[44] Church, H.B. Curry, and Rosser were among those on the committee that founded the association; and Church became the principal editor of the journal, a position he held for many years at Princeton and later at UCLA. Lack of funding placed the journal in jeopardy in its first year, but Dean Gauss of the Princeton Graduate School organized a plan for research institutions to provide subvention funds until the journal could survive on the association's membership dues. This plan was carried out successfully, and Princeton University and the Institute for Advanced Study were two of the nine institutions that contributed. Church's involvement was crucial in giving the journal the research importance it has today.

Church and Veblen shared a strong interest in developing symbolic logic as a branch of mathematics rather than as a branch of philosophy. Church wrote to Veblen in 1935 at his summer home in Maine to report the "unfortunate" choice of C.H. Langford as his co-editor for the journal, and of the possibility that philosopher C.I. Lewis might decline the presidency of the association.[45] Church wanted to promote Hermann Weyl for president instead of Lewis, but was hampered by Weyl's unwillingness to support a journal or association of logic proposed by philosophers. Church continued:

> Weyl is surely mistaken in supposing that there is no need for the proposed journal. Mathematical journals accept papers in symbolic logic with considerable reluctance, and clearly do not have the space to publish all that deserves publication, as I know from experience, not only with my own papers and those of Rosser and Kleene, but in refereeing papers of others. In regard to Weyl's apparent objection to the part of philosophers, which is probably based on the admittedly considerable amount of nonsense published by philosophers in this field, it seems to me that the sensible thing is to try to direct the present undertaking into worthwhile channels, rather than to ignore it or try to suppress it. It is precisely for this reason that I would like to see Weyl, or someone else of genuine ability and understanding, as president.
>
> Of course, when you get right down to it, mathematicians also have been known to publish nonsense in the name of symbolic logic.
>
> Anywhere [sic], there is undoubtedly a need for the advent of a younger generation of philosophers, especially philosophical logicians, with a really adequate understanding of mathematics. And a joint enterprise between mathematicians and philosophers like the present one would seem to be a good way of contributing towards such a result.[46]

Veblen should be remembered particularly for his role in shaping the careers of Alonzo Church and Kurt Gödel. Veblen singled out Church from among his undergraduate students and encouraged him to remain at Princeton for graduate study. Veblen arranged for Church's fellowship support as a graduate student, introduced him to the study of logic, and eventually directed his dissertation on the axiom of choice. Church remembers being attracted to Veblen's courses because of Veblen's interest in foundational issues. As

Church writes:

> I was generally interested in things of a fundamental nature. As an under-graduate even I published a minor paper about the Lorentz transformation, the foundations of relativity theory. It was partly through this general interest and partly through Veblen, who was still interested in the informal study of foundations of mathematics. It was Veblen who urged me to study Hilbert's work on the plea, which may or may not have been fully correct, that he himself did not understand it and he wished me to explain it to him. At any rate, I tried reading Hilbert. Only his papers published in mathematical periodicals were available at that time. Anybody who has tried those knows they are very hard reading. I did not read as much of them as I should have, but at least I got started that way. Veblen was interested in the independence of the axiom of choice, and my dissertation was about that. It investigated the consequences of studying the second number class, that is, subordinates, if you assume certain axioms contradicted the axiom of choice.[47]

This kind of interest in logic and foundations was not shared by other members of the Princeton mathematics faculty in the 1920s, or by anyone in the philosophy department. As Church said of the mathematics faculty's interest in logic in the 1930s after he had joined their ranks, "There were not many others interested in this field, and it was not thought of as a respectable field, with some justice. There was a lot of nonsense published under this heading."[48]

While Church was off in Göttingen in 1929 on the second year of a National Research Council post-doctoral fellowship, Veblen convinced his colleagues to hire Church as an assistant professor.[49] Veblen continued to follow Church's career with interest and to support his advancement through the professorial ranks.

Veblen's role in bringing Kurt Gödel to the institute is less clear, but certainly important.[50] It is suspected that Veblen learned of Gödel's work through his colleague John von Neumann, who had made fundamental contributions to set theory and proof theory in the 1920s, and was one of the first people to appreciate the significance of Gödel's incompleteness results. Gödel made three visits to the institute between 1933 and 1939. In 1939, with the Nazi administration interfering with his efforts to obtain a teaching position in Vienna and threatening his conscription to military service, Gödel wrote to Veblen for help with emigration.[51] Somehow, the necessary visas and exit permits were arranged, and Gödel arrived in Princeton to become an ordinary member of the institute. Gödel remained at the institute the rest of his career.[52]

Conclusions

Veblen was an able research mathematician with an abiding interest in geometry and its foundations. Somewhat by accident, these interests led him to logic. Following Hilbert's lead, Veblen believed the only secure foundation for geometry was an axiomatic foundation; and to demonstrate the adequacy of his axiomatic systems he found that he needed to hone the tools of logic. In so doing, he made some minor contributions to the technical field of mathematical logic, as we know it today, as well as more significant applications of logic to the foundations of geometry.

These studies engendered in Veblen a general interest in logic, one that was not shared by many of his mathematical colleagues in Princeton or elsewhere. He believed there should be a close relationship between logic and mathematics. Logic was a subject that should be developed for the service it could provide to mathematics; and to serve well, logic should be developed into a mathematical discipline. He inculcated these beliefs in the graduate students he trained at Princeton, in particular into Alonzo Church. And these beliefs were realized through the work of Church and Gödel, mathematicians whose careers he advanced in important ways.

Largely through the stage-setting efforts of Veblen, Princeton became a leading national and international research center in logic. This came about not so much because of Veblen's interest in logic, as through the strong mathematical research programs he helped to create, first at Princeton University and later at the Institute for Advanced Study. The general strength in Princeton mathematics was critical to the rise of logic there. Veblen, perhaps better than any other mathematician of his time, understood the importance of environmental factors to the growth of mathematics. And it was these environmental factors that caused logic to flourish in Princeton.

Notes

1. Leading European centers of logic included Amsterdam, Cambridge, Göttingen, Turin, and Warsaw—though none of them maintained their strength in logic throughout the entire half century under consideration.

2. Further evidence of the weakness of American logic is given by prominent examples of aspiring American logicians being encouraged to complete their logic education in Europe: Veblen sent Alonzo Church to Göttingen for a year; Norbert Wiener visited England to study with Bertrand Russell. This is part of a more general pattern of European study for young Americans aspiring to mathematical research careers, which was accomplished regularly after 1924 with the assistance of National Research Council post-doctoral fellowships. The number of Americans who pilgrimaged to Europe is uncertain; and some Americans, such as Emil Post and the students of C.S. Pierce at Johns Hopkins, never felt a need for European study. A better picture will only emerge after a systematic study of the American logical community, e.g., an examination of the educations of the early members of the Association for Symbolic Logic.

3. There is little written about the history of American logic. Birkhoff presents in the American Mathematical Society *Semicentennial Addresses* (1938) a general overview in a seven-page subsection on "Symbolic Logic and Axiomatics." Van Evra (1966) contains historical descriptions of the work of Alonzo Church, C.I. Lewis, C.S. Pierce, and W.V.O. Quine. Dipert (1978) provides historical insight into Pierce's

work. Scanlan (1982) contains material on the American Postulate Theorists. There is a need for a history of this subject, which would cover both the technical results and their social context (development of research faculties and doctoral programs, professional societies and journals, funding, etc.). A project, directed by Christian Thiel, on "Sozialgeschichte der Logik" is now underway. Thony Christie and Karin Beikuefner, part of the research team, are covering aspects of American logic. See Peckhaus (1986) for details.

4. These names are listed in chronological order of the date the Ph.D. was awarded, from Church in 1927 to Kemeny in 1949. The Princeton experiences of some of these logicians are given in the Princeton Mathematics Community (PMC) in the 1930s Oral History Project, the results of which are held by the Mudd Library, Princeton University. In particular. see the comments of Alonzo Church (PMC 5), Alfred Leon Foster (PMC 12), Leon Henkin (PMC 14, 19), John Kemeny (PMC 22), Stephen Kleene (PMC 23), and J. Barkley Rosser (PMC 23). Secondary copies of the transcripts are held by repositories around the United States, including the Charles Babbage Institute of the University of Minnesota.

5. See Davis (1965) and Kleene (1981) on the history of recursive function theory. The author is also preparing an article on the antecedents of this development. See Kleene (1987) and Feferman et al. (1986) on the life and work of Gödel. John Dawson is preparing a full-length biography of Gödel.

6. For biographical information about Veblen see Archibald (1938, pp. 206–211), MacLane (1976), and especially Montgomery (1963). Veblen was such a central figure in the Princeton mathematics community that he is discussed in 35 of the 45 interviews in the Princeton Mathematics Community in the 1930s Oral History Project (Mudd Library, Princeton University). See, especially, the interviews with Albert Tucker.

7. Written and delivered at a meeting of the American Mathematical Society in 1901, Moore (1902) presents projective axioms (axioms of connection and order) for the general n-dimensional geometry.

8. Moore's seminar also inaugurated the research program in foundations of geometry of Nels J. Lennes, who occasionally worked jointly with Veblen on this subject. R.L. Moore entered the graduate program at Chicago in 1903 and became interested in the foundations of geometry under the influence of E.H. Moore and Veblen. See Johnson (1984) for additional information.

9. Veblen (1904a).

10. See Pasch (1882), Peano (1889; 1894), Hilbert (1899), and Pieri (1899).

11. Tarski was the first to note, in 1934, that Veblen had made a logical error. See Tarski (1956a, pp. 306–307; 1956b). Wilder (1965, p. 41) describes the error as follows: "Or consider the system of Euclidean geometry given by Veblen in 1904[a]. Only two terms, 'point' and 'order,' were left undefined, all other geometrical terms presumably being defined in terms of these. However, as was first pointed out by Tarski ... the definition of 'congruent' involves not just 'point' and 'order,' but a geometric entity E which is not postulated by the system, but is arbitrarily chosen. Consequently, it is possible to find two models of Veblen's system consisting of the same points and having the same order relations between points, but in which congruence is different in the sense that two given pairs of points may be congruent in one model but not in the other."

12. On the difference between definitions and axioms, Veblen points to Padoa (1901), E.H. Moore (1903), and Veblen (1903). There seems to be an implicit criticism here of Hilbert's logical indiscretions in the *Grundlagen der Geometrie*.

13. Veblen notes that the term "categorical" was introduced by John Dewey, and that Hilbert (1899) and Huntington (1902) had previously investigated the categoricity of systems. Corcoran (1980; 1981) points out that Veblen infers completeness from categoricity without argument, and apparently does not distinguish between the two concepts. Moore (1988, Section 6) also provides a detailed historical account of categoricity.

14. Veblen also wrote two popular articles (1905a, 1906) in this period on the foundations of Euclidean geometry, which provide historical accounts of the search for a logically rigorous axiomatic foundation. He continued to write about the foundations of geometry later in his career. Veblen (1911) presents a popular account of the modern axiomatic treatment of Euclidean geometry.

15. Veblen (1905b). Veblen cites Cantor (1895) and Huntington (1905) as a foundation for his work.

16. The Jordan Curve Theorem is proved in Veblen (1905c); the studies of compactness are made in Veblen (1904b) and Veblen and Lennes (1907).

17. Veblen and Lennes (1907, p. iv). On page iii the authors write, "The general aim has been to obtain rigor of logic with a minimum of elaborate machinery. It is hoped that the systematic use of the Heine–Borel theorem has helped materially toward this end...."

18. Veblen and Young (1908). They were also careful to show the system was categorical.

19. Veblen and Young (1910/1918). As the introduction to the second volume indicates, Veblen wrote the second volume largely without assistance from Young.

20. Their interest in the abstract group classification of geometries is discussed in the Preface to volume II.

21. Veblen and Young (1918, p. iv).

22. Veblen and Young (1910, pp. iii–iv).

23. Veblen's 1916 Cambridge Lectures, published as Veblen (1922), stood for many years as a standard source on topology; Veblen and Whitehead (1932) had similar stature for the study of differential geometry.

24. Veblen and Whitehead (1931). Characteristic of Veblen, the authors were careful to demonstrate the consistency and independence of the axioms. See also the account in Veblen and Whitehead (1932, Ch. 6).

25. Veblen (1932, p. 17). The first of Veblen's three Rice Institute lectures, entitled "The Modern Approach to Elementary Geometry" and published in Veblen (1934), presents a beautifully drawn summary of Veblen's mature views on geometry and its foundations.

26. Published as Veblen (1925). In his address he also remarked on the relationship between the mathematical and physical investigations of space, e.g. between Euclidean geometry and Newtonian physics, and between Riemannian geometry and relativity theory—and what this means for the foundations of geometry. In a 1922 address to the Mathematics Section of the American Academy for the Advancement of Science, published as Veblen (1923), he acknowledges the value that the abstract, i.e., axiomatic, approach has had for geometry, and provides some refections on the value of this to the study of classical and modern mechanics and other branches of physics.

27. Veblen (1925, p. 140).

28. Veblen (1925, p. 141).

29. The story of the rise of Princeton mathematics is told in greater detail in Aspray (1988). Also see the transcripts of the Princeton Mathematics Community in the 1930s Oral History Project.

30. Letter, Veblen to Vernon Kellogg, 10 June 1924. Oswald Veblen Papers, Library of Congress.

31. Ibid.

32. Anonymous (1931, p. 113). In a letter, 19 February 1929, Veblen describes to Luther Eisenhart the rooms at Burlington House, home of the Royal Society, and Cambridge as models for Princeton. He also quotes from a letter he had received from their benefactor, Thomas Jones, dated 25 January 1929: "the building is not to 'be a laboratory in which a great deal of manual work as well as brain work must be carried on; it should be a house of mathematics—a haunt of mathematicians in which the spirit and the traditions of generations will be perpetuated and cherished. That cannot be accomplished in a barn'." Fine Hall was certainly not a barn. Veblen mentioned elsewhere using the universities of Paris and Oxford as models for Fine Hall.

33. Veblen desired to promote close ties between mathematics and physics; so Fine Hall was placed adjacent to Palmer Laboratory of Physics and a connecting corridor was built, the holder of the chair in mathematical physics was given an office in Fine Hall, the library contained both the mathematics and physics research literature, and the physicists were welcomed to afternoon tea.

34. Anonymous (1931, p. 112).

35. Bienen (1970, p. 17).

36. See Flexner (1960).

37. Private communication from Albert W. Tucker, 1985.

38. There is some correspondence in the Veblen Papers (Library of Congress) with Eisenhart over the potential damage the institute will cause the university mathematics program. See especially Eisenhart's letter of 15 August 1932.

39. For example, Richard Brauer and Nathan Jacobson were assistants to Weyl; Banesh Hoffmann and Leopold Infeld were assistants to Einstein; Wallace Givens, Abraham Taub, and J.H.C. Whitehead were assistants to Veblen; Henry Wallman was assistant to Alexander; and William Duren and Everett Pitcher were assistants to Morse. The only IAS faculty member to use his assistants to assist with team research projects was Morse.

40. Since dissertation direction by institute faculty was informal, it is difficult to verify. However, Albert Tucker has told me he believes that the dissertations of Wallace Givens, Israel Halperin, and Joachim Weyl (Hermann Weyl's son) were informally directed by Veblen, von Neumann, and Weyl, respectively. (Private communication, 4 January 88).

41. A remark made in 1936 before an international scientific audience. As quoted in Chapin (1958).

42. See, for example, Rosser's comments in the oral history the author conducted with Rosser and Stephen Kleene, 26 April 1984. PMC 23, the Princeton Mathematics Community in the 1930s Oral History Project.

43. By which he meant Rosser, Kleene, Church, and Curry. The other fifth, Quine, was at Harvard. (Rosser/Kleene interview, op. cit.)

44. See Ducasse and Curry (1962) for additional information about the early history of the association and the journal.

45. This and the remaining information in the paragraph is in a letter from Church to Veblen, 19 June 1935 (Oswald Veblen Papers, Library of Congress). In an earlier draft of this paper the author interpreted this letter as evidence that Church was biased against philosophers, to which one knowledgeable reviewer took strong exception. The remark about Langford in the letter is inconclusive; there may have been personal or other reasons besides disciplinary status that cause Church to view Langford's

appointment as "unfortunate." In the author's interview with Church (see footnote 47), however, Church characterizes Lewis as a "leader" among "philosophers who were interested in symbolic logic frcm the point of view of its relevance to philosophy rather than math" and Church indicated his unwillingness to use Lewis's logic text for this reason. I do believe that Lewis's philosophical orientation was the reason Church favored Weyl's appointment as president.

46. Ibid.

47. Alonzo Church, Oral History PMC 5, 17 May 1984, conducted by the author. The Princeton Mathematics Community in the 1930s Oral History Project.

48. Church interview, op. cit.

49. The Veblen Papers (Library of Congress) include a regular correspondence between Veblen and Church during the two years Church was away from Princeton on a National Research Council fellowship. They correspond about the work of Brouwer, Curry, Hilbert, Ramsey, and Russell (see, especially, Veblen's letter of 28 April 1928 about Hilbert's continuum paper); who Veblen recommends Church study with during the second year of his fellowship (Brouwer, Hilbert, or Weyl); how to help Curry get a fellowship and teaching position; and information on the new mathematics building at Göttingen (Church sent floor plans on 26 January 1929).

50. This account is taken largely from Feferman et al. (1986), which includes additional biographical information about Gödel and about his contributions while at the institute.

51. There were at least two reasons for Gödel to have written Veblen about this matter: Gödel was scheduled to visit the institute in 1939, and Veblen handled many such personnel matters (there being almost no professional administration at the institute); also, Veblen was a leader in the United States in aiding refugee mathematicians, helping dozens of mathematicians to emigrate and avoid the Nazi peril. Gödel had notoriously frail physical and mental health and knew the deleterious effect military service would have.

52. Gödel was made a permanent member of the institute in 1946 and professor in 1953.

Bibliography

American Mathematical Society (1938), *Semicentennial Addresses*, American Mathematical Society, New York. (Volume 2 of American Mathematical Society Semicentennial Publications.)

Anonymous (1931), A memorial to a scholar–teacher, *Princeton Alumni Weekly*, 30 October 1931, pp. 111–113.

Archibald, R.C. (1938), *A Semicentennial History of the American Mathematical Society 1888–1938*, Vol 1, American Mathematical Society, New York.

Aspray, W. (1988), The emergence of Princeton as a world center for mathematical research, 1896–1939, in W. Aspray and P. Kitcher (eds.), *History and Philosophy of Modern Mathematics*, University of Minnesota Press, Minneapolis, pp. 344–364.

Bienen, L.B. (1970), Notes found in a Klein bottle, *Princeton Alumni Weekly*, 21 April 1970, pp. 17–20.

Cantor, G. (1895), Zur Begründung der transfiniten Mengenlehre, I, *Math. Ann.* **46**, 481–507.

Chaplin, V. (1958), A history of mathematics at Princeton, *Princeton Alumni Weekly*, 9 May 1958.

Corcoran, J. (1980), Categoricity, *Hist. Philos. Logic* **1**, 187–207.

Corcoran, J. (1981), From categoricity to completeness, *Hist. Philos. Logic* **2**, 113–119.

Corcoran, J. (1986), Undefinability tests and the Erlanger program: Historical footnotes, *Proceedings, International Congress of Mathematicians, Berkeley 1986*, p. 339.

Davis, M. (1965), *The Undecidable: Basic Papers on Undecidable Propositions, Unsolvable Propositions, and Computable Functions*, Raven Press, Hewlett, NY.

Dipert, R. (1978), Development and Crisis in Late Boolean Logic: The Deductive Logics of Peirce, Jevons, and Schröder, Ph.D. Dissertation, Indiana University.

Ducasse, C.J. and Curry, H.B. (1963), Early history of the Association for Symbolic Logic. *J. Symbolic Logic* **27**, 255–258. Addendum: *J. Symbolic Logic* **28**, p. 279.

Eisenhart, L.P. (1931), The progress of science: Henry Burchard Fine and the Fine Memorial Hall, *Scientific Monthly* **33**, 565–568.

Feferman, S. et al. (eds.) (1986), *Kurt Gödel Collected Works*, Vol. I, Oxford University Press, New York. (*Gödel's Life and Work*, pp. 1–36.)

Fine, H.B. (unsigned) (1926), *The Role of Mathematics*, The Princeton Fund Committee.

Flexner, A. (1960), *Abraham Flexner: An Autobiography*, Simon and Schuster, New York.

Hilbert, D. (1899), *Grundlagen der Geometrie*, Teubner, Leipzig. English trans. by E.J. Townsend, Open Court, Chicago, 1902.

Hille, E. (1962), *In Retrospect*, Notes, Yale Mathematical Colloquium, 16 May 1962.

Huntington, E.V. (1902), A complete set of postulates for the theory of absolute continuous magnitude, *Trans. Amer. Math. Soc.* **3**, 264–279.

Huntington, E.V. (1905), A set of postulates for real algebra, comprising postulates for a one-dimensional continuum and for the theory of groups, *Trans. Amer. Math. Soc.* **56**, 17–41.

Johnson, D. (1984), Mathematical Research in the United States from 1890: The Case of the University of Chicago, Hampden–Sydney College, unpublished.

Kleene, S.C. (1981), Origins of recursive function theory, *Ann. Hist. Comput.* **3**, 52–67.

Kleene, S.C. (1987), Kurt Gödel, *Biographical Memoirs Nat. Acad. Sci.* **56**, 135–178.

MacLane, S. (1976), Oswald Veblen, *Dictionary of Scientific Biography*, 599–600.

Montgomery, D. (1963), Oswald Veblen, *Bull. Amer. Math. Soc.* **69**, 26–36.

Moore, E.H. (1902), On the projective axioms of geometry, *Trans. Amer. Math. Soc.* **3**, 142–158.

Moore, E.H. (1903), On the foundations of mathematics, *Bull. Amer. Math. Soc.* (2), **9**, 402–424.

Moore, G.H. (1988), The emergence of first-order logic, in W. Aspray and P. Kitcher (eds.), *History and Philosophy of Modern Mathematics*, University of Minnesota Press, Minneapolis, pp. 95–135.

Padoa, A. (1901), Logique et Histoire des Sciences, *Bibliothèque du Congrès International de Philosophie*, III.

Pasch, M. (1882), *Vorlesungen über neuere Geometrie*, Teubner, Leipzig.

Peano, G. (1889), *I principii di geometria*, Fratelli Bocca, Turin.

Peano, G. (1894), Sui fondamenti della geometria, *Riv. Mat.* **4**, 51–59.

Peckhaus, V. (1986), Case studies towards the establishment of a social history of logic, *Hist. Philos. Logic* **7**, 185–186.

Pieri, M. (1899), Della geometria elementare come sistema ipotetico deduttivo. Monografia del punto e del moto, *Memorie della Reale Accademia delle Scienze di Torino* (2), **49**, 173–222.

Reingold, N. (1981), Refugee mathematicians in the United States of America, 1933–1941: Reception and reaction, *Ann. Sci.* **38**, 313–338.

Scanlan, M. (1982), American Postulate Theorists and the Development of Axiomatic Method, 1901–1930. Ph.D. Dissertation, SUNY, Buffalo.

Tarski, A. (1956a), *Logic, Semantics, Metamathematics: Papers from 1923 to 1938*, trans. by J.H. Woodger. Clarendon, Oxford.

Tarski, A. (1956b), A general theorem concerning primitive notions of Euclidean geometry, *Indag. Math.* **18**, 468–474.

Van Evra, J. (1966), A History of Some Aspects of the Theory of Logic, 1850–Present. Ph.D. Dissertation, Michigan State University.

Veblen, O. (1903), Hilbert's foundations of geometry, *Monist*, **13**, 303–309.

Veblen, O. (1904a), A system of axioms for geometry, *Trans. Amer. Math. Soc.* **5**, 343–384.

Veblen, O. (1904b), The Heine–Borel theorem, *Bull. Amer. Math. Soc.* **10**, 436–439.

Veblen, O. (1905a), Euclid's parallel postulate, *Open Court* **19**, 752–755.

Veblen, O. (1905b), Definition in terms of order alone in the linear continuum and in well-ordered sets, *Trans. Amer. Math. Soc.* **6**, 165–171.

Veblen, O. (1905c), Theory of plane curves in nonmetrical analysis situs, *Trans. Amer. Math. Soc.* **6**, 83–98.

Veblen, O. (1906), The foundations of geometry: An historical sketch and a simple example, *Popular Science Monthly* n.v., pp. 21–28.

Veblen, O. (1911), The foundations of geometry, in J.W.A. Young, *Mongraphs on Topics of Modern Mathematics*, Longmans, Green, New York, Chapter I, pp. 1–51.

Veblen, O. (1922), *Analysis Situs*, American Mathematical Society Colloquium Publications, volume 5, part 2.

Veblen, O. (1923), Geometry and physics, *Science* **57**, 129–139.

Veblen, O. (1925), Remarks on the foundations of geometry, *Bull Amer. Math. Soc.* **31**, 121–141.

Veblen, O. (1934), Certain aspects of modern geometry—A course of three lectures, *Rice Institute Pamphlets* **31**, 207–255.

Veblen, O. and Alexander, J.W. (1913), Manifolds of n dimensions, *Acta Math.*, Series 2, **14**, 163–178.

Veblen, O. and Lennes, N.J. (1907), *Introduction to Infinitesimal Analysis*, Wiley, New York.

Veblen, O. and Whitehead, J.H.C. (1931), A set of axioms for differential goemetry, *Proc. Nat. Acad. Sci.* **17**, 551–561.

Veblen, O. and Whitehead, J.H.C. (1932), *Foundations of Differential Geometry*, Cambridge Tracts in Mathematics and Mathematical Physics, number 29, Cambridge University Press, Cambridge.

Veblen, O. and Young, J.W. (1908), A set of assumptions for projective geometry, *Amer. J. Math.* **30**, 347–380.

Veblen, O. and Young, J.W. (1910/1918), *Projective Geometry*, 2 volumes (volume two by Veblen), Ginn, Boston.

Wilder, R.L. (1965), *Introduction to The Foundations of Mathematics*, Wiley, New York.

The Löwenheim–Skolem Theorem, Theories of Quantification, and Proof Theory*

Irving H. Anellis

Warren Goldfarb (1971, p. 17) wrote that "Herbrand's work had an immediate impact on the Hilbert school," and quotes Paul Bernays (Hilbert and Bernays, 1934, 1939, vol. 1, "Foreword") to the effect that "the appearance of the works of Herbrand and Gödel have altered proof theory." We are concerned here to explore the role which Herbrand's work had in developing the discipline of proof theory. More specifically, we are interested in formulating a theory about the impact which questions raised by Herbrand, about the meaning of *satisfiability* in Hilbert's axiomatic method, had on the proliferation of quantification theories presented as alternatives to Hilbert's system.

Important surveys of proof theory research have been made by Kreisel (1968, 1971), Minc (1975), and Prawitz (1971). Van Heijenoort (1976) has studied the development in theories of quantification, and Goldfarb (1979) has sought to present an historically accurate understanding of the way in which logicians such as Frege, Russell, Hilbert, and Herbrand, among others, understood their own contributions to quantification theory in the context of the interpretation of quantifiers for a universe of discourse. Thus, he is interested in the relations between syntax and semantic interpretation on one hand, and deducibility under quantifier interpretations for each formal system on the other.

In what follows, I wish to argue that the impetus for the development of proof theory, and the competing quantification theories developed after 1931, is to be sought in the questions raised by Herbrand (1930a) in the attempt to clarify the concept of *being a proof* for a Hilbert-type quantification theory.

---

* An early version of this chapter was presented to the American Mathematical Society on 25 January 1979 (Anellis, 1979), under the title *The Löwenheim–Skolem theorem, theories of quantification, and Beweistheorie.* Under that same title, it was to have appeared in the planned but never-executed *Proof Theory, Constructive Mathematics and Applications: Proceedings of the American Mathematical Society Special Session on Proof Theory*, to have been edited by this author and G.E. Minc.

The current version of the chapter has been improved by the helpful comments of G.E. Minc and Thomas Drucker.

The Hilbert system, along with related systems, those of Frege and of Whitehead and Russell, precede the work of Herbrand. By contrast, the important work of Jaśkowski (1934) and Gentzen (1934) followed three years after the work of Herbrand. Moreover, we must note that Gödel's (1931) incompleteness results are contemporary with Herbrand's thesis. Axiomatic systems, of course, began with Frege's *Begriffsschrift* (1879) and reached their peak in the *Principia Mathematica* of Whitehead and Russell (1910–1913). By 1931, there were two major axiomatic systems—the Frege-type system and the Hilbert-type system. Here I wish to examine the connections between proof theory and theories of quantification, and to explain how questions about the nature of proofs for axiomatic systems led to the development of alternatives to those systems.

The historical evidence available suggests that Herbrand's thesis (1930) and Gödel's incompleteness results (1931), appearing within months of one another, together had a profound impact on the primacy of the axiomatic method. Certainly, the Russell paradox and the theory of types introduced into set theory to resolve the paradox can be seen as the best example known of the incompleteness of the Whitehead–Russell axiomatic system as defined by Gödel's theorem. The examination by Herbrand (1930a) of the meaning of *being a proof* for the Hilbert system forced serious questions to be asked about Hilbert-type systems. We are interested here in the technical repercussions of the work of Herbrand, and argue that these repercussions of Herbrand's study of the nature of proof in Hilbert's system included the development of alternative theories of quantification. It is not coincidental that the majority of alternative formal quantification theories devised to rival the axiomatic method arose within a few years of the publication of both the Gödel results and, more particularly, Herbrand's study.

It is not clear that there is an historical relationship between Gödel's theorem and Herbrand's study of Hilbert's concept of *proof*. For van Heijenoort, the role of Gödel's incompleteness theorem is, if not central, then certainly crucial, in the history of logic, and played a significant role not only in the history of metamathematics, but also in the development of philosophies of mathematics. My investigation indicates that the technical developments in the Hilbert-type systems and the rise of the other systems were stimulated less by the Gödel incompleteness results than by questions raised, through the studies of Herbrand, by the Löwenheim–Skolem theorem on $\aleph_0$-satisfiability. Herbrand's (1930a) efforts, and his use of the Löwenheim–Skolem theorem in that study, are central to the further development of quantification theory and provide the crucial link between proof theory and theories of quantification.

Kreisel (1976, p. 128) has said that, "though the aims of Hilbert's program and the theory of proofs are usually unrelated and often in conflict, it should not be assumed that the aims of that theory are in conflict with Hilbert's own interests," and that "at least at one time, his principal interest was to attract the attention of mathematics to the topic of *mathematical proofs as an object of* (mathematical) *study*: and the 'grand' program was to have been the bait."

If we understand by quantification theory that fragment of Frege's or Russell's system that contains no variables of type greater than 2 and no quantifiers but those taking only individuals within their scope—i.e., first-order functional calculus (with identity)—then, says Goldfarb (1979, pp. 352–353), "we cannot expect quantification theory to arise naturally within the logicist program, because neither Frege nor Russell treated metamathematical, i.e., proof-theoretic, questions about the *underlying logic* of the various areas of mathematics." This claim of Goldfarb's is difficult to accept: surely both Frege and Russell aimed specifically to develop their systems for the logicization of mathematics (see, e.g., Frege's *Über den Zweck der Begriffsschrift* (1883). It would clearly be a false claim if applied to Hilbert's program. Hilbert (1900) is precisely an application of the study of the logic of proofs to the formalization of the real number system, and hence to analysis. The correct interpretation is that quantification theory arose as part of the attempt to clarify Hilbert's conception of the nature of proofs. And this was done by Herbrand. Herbrand used the opportunity provided in his investigation of Hilbert's system to set forth the first alternative quantification theory, namely, Herbrand quantification (Q_H), the Gentzen sequent calculus, and the method of natural deduction introduced by Jaśkowski appeared soon after, in 1934, as rivals to the axiomatic method and to Q_H.

According to Jean van Heijenoort (1976, p. 7), the theory of quantification is a "family of formal systems" whose members include Hilbert-type, or axiomatic, systems, Herbrand Quantification, the Gentzen Sequent Calculus, and Natural Deduction. Van Heijenoort (1976) traces the history of the evolution or development (*desarrollo*) of quantification theory, starting from the definition of the Hilbert program (see also Kreisel 1959, p. 150), according to which "*proof* itself is an object of study" for mathematics) and the Herbrand system, through the Gödel incompleteness results and consideration of the Gentzen Sequent Calculus and Natural Deduction. Finally, he ties them all together for comparison and for enumeration of each of their respective contributions towards clarification of the concept of *proof*.

It is clear from Herbrand's own comments that his investigations were undertaken to clarify the concept of *being a proof* for a Hilbert-type quantification system. In the introduction to his *Recherches sur la théories de la démonstration* (1930a), he speaks of the recursive method to "prove that every true proposition of a given theory has a property A" (see the translation by Goldfarb (1971, p. 49)), and immediately ties this to the finitist limit on recursive proofs enunciated by Hilbert. For Herbrand, this finitist limit demands a challenge specifically to the transfinitist proofs of Löwenheim (in terms of $\aleph_0$-satisfiability) and require that Löwenheim's infinite conjunction be reinterpreted as Herbrand expansion. All of this gives impetus to investigations in proof theory and into the concept of *validity*. The Gödel results did not serve to initiate these investigations. They were already underway in Herbrand's work, months before Gödel announced his results. The evidence is that much of Herbrand's concern centered explicitly on the Löwenheim–

Skolem theorem, and particularly on the idea of $\aleph_0$-satisfiability obtained from the k-satisfiability for every finite k of Löwenheim infinite conjunction. Herbrand (1930b) dealt separately with Hilbertian finitism.

The alternatives to the axiomatic method cam fast and furious after 1931; but the example of Herbrand shows that the Gödel results did not necessarily precede—or cause—the development of alternative quantification theories. We can accept the claims of van Heijenoort (1976) that:

(1) theory of quantification is a "family of formal systems";
(2) the several theories of quantification, rather than being in competition, represent a natural development (*desarrollo*).

However, to this set of conclusions must be added the thesis that:

(3) technical developments in Hilbert-type systems, including development of *Beweistheorie* by Hilbert–Bernays (1934, 1939), and development of alternative theories of quantification, are primarily due to questions raised by the Löwenheim–Skolem theorem rather than by Gödel's incompleteness results.

The axiomatic method provides results on the basis of the concept of *formal system*. We are provided with a set of *axioms*, a list of formulas from which we derive other formulas, the *theorems* of the system, and an *inference rule*, the rule of detachment, by which theorems may be derived from the axioms of the system. The concept of *proof* is never considered; it is extrasystematic, metamathematical. Löwenheim (1915), for example, had no notion of *proof*, but only (as van Heijenoort, (1977, p. 183) reminds us) of *satisfiability* and *validity*, whereas Skolem was looking for a proof procedure (for completeness) (see van Heijenoort (1977, p. 185)). A *proof* is just a string or sequence of formulas or axioms, the last line of which, the *Endformula*, is a theorem proved in the system. (The Hilbert-type axiomatic system is defined by the Hilbert program with its underlying formalist philosophy (see, e.g., Hilbert (1918); also Kreisel (1959).)

Herbrand, following the Löwenheim–Skolem results on satisfiability, was led to seek a clarification of satisfiability or validity. This proof-theoretic investigation began with the Löwenheim–Skolem theorem which, in its simplest form, states that if full first-order quantification theory Q (i.e., first-order functional calculus with identity) has an interpretation in a subdomain Q' of Q, then Q has a countable model, or is satisfiable. More precisely; if Q is satisfiable for some finite k, then it is $\aleph_0$-satisfiable. And any class K having a formula P in Q which is k-satisfiable for every countable k is $\aleph_0$-satisfiable, or has a formula P' in Q which is a copy of P, and P' is $\aleph_0$-satisfiable. As stated by Skolem (1923) it reads:

> If [a set of] axioms are consistent, then there is a domain B for which the axioms are satisfied, and for which all elements of B are enumerable with the use of the positive finite integers.

We may reformulate this in our terminology as follows:

Theorem 1 (Löwenheim–Skolem). *If full first-order quantification theory Q is satisfiable for some finite k, then it is $\aleph_0$-satisfiable; and any class K having a formula F in Q which is k-satisfiable for every countable k is $\aleph_0$-satisfiable, or has a formula F' in Q which is a copy of F, and F' is $\aleph_0$-satisfiable.*

Löwenheim (1915) showed that if a finitely valid formula F is not valid, then it is not $\aleph_0$-valid; or by contraposition, that if F is $\aleph_0$-valid, then it is k-valid for any countable k, where it is not invalid for some finite domain. Skolem (1920) proved this theorem and generalized it to a denumerably infinite set of formulas, such that:

(a) if F is satisfiable, then F is satisfiable in the domain of natural numbers, where a proper assignment is made to the predicate letters of F; and

(b) if F is satisfiable, i.e., there exists a domain D in which F is satisfiable, then F is satisfiable in a denumerable subdomain D' of D. for the same assignment of meanings to predicate letters of F.

Skolem (1923) was able to simplify his proof by utilizing the law of infinite conjunction and dispensing with the Axiom of Choice used in the earlier proof. On the basis of the infinite conjunction used by Skolem to obtain his results on $\aleph_0$-satisfiability in Theorem 1, Herbrand (1930a) was able to provide an expansion, now known as *Herbrand expansion*, and generate his *Théorème Fondamental* (see Goldfarb (1971, pp. 168–169) and van Heijenoort (1968, p. 554)).

"Herbrand's work can be viewed," says van Heijenoort (1968 Editor's Introduction, p. 526) "as a reinterpretation, from the point of view of Hilbert's program, of the results of Löwenheim and Skolem. Of his fundamental theorem Herbrand (1931, p. 4) writes "'that it is a more precise statement of the Löwenheim–Skolem theorem'."

In general form, we have:

Theorem 2 (Fundamental Theorem, Herbrand). *For some formula F of classical quantification theory, an infinite sequence of quantifier-free formulas F_1, F_2, ... can be effectively generated, for F provable in (any standard) quantification theory, if and only if there exists a k such that F is (sententially) valid; and a proof of F can be obtained from F_k.*

As originally stated by Herbrand, this says:

(1) *If for some number p a proposition P has property B of order p, the proposition P is true. Once we know any such p, we can construct a proof of P.*

(2) *If we have a true proposition P, together with a proof of P. we can, from this proof, derive a number p such that P has property B of order p.*

Let us give an example. To do so, we require some definitions. A formula of Q_H is called *rectified* if it contains no vacuous quantifier, no variable has free or bound occurrences in it, and no two quantifiers bind occurrences of the same variable. A quantifier in a rectified formula is called *existentialoid* if it is existential and in the scope of an even number of negation symbols, or universal and in the scope of an odd number of negation symbols; otherwise it is *universaloid*. A variable in a rectified formula is an *existentialoid variable* if and only if it is bound by an existentialoid quantifier, and is *universaloid variable* if and only if it is bound by a universaloid quantifier (see van Heijenoort (1975, p. 1) and (1982, p. 58)).

EXAMPLE. If $\mathrm{Exp}[F, D_k]$ is a Herbrand expansion of F over D_k (where D_k is the kth element of the sequence D containing the individual constants and universaloid variables of F and the elements of $D_0, \ldots, D_{k-1}$, where initial D_0 contains the individual constants and the existentialloidally free universaloid variable F, and where, if F contains none of these, the one element of D_0 is an arbitrary individual contant u_0), then the formula F is provable in classical quantification theory if and only if there exists a k such that $\mathrm{Exp}[F, D_k]$ is sententially valid.

Thus, for F_Q a formula of classical quantification theory without identity, for the kth Herbrand expansion $F_{Q_{H_k}}$ for any k, $F_Q \rightarrow F_{Q_{H_k}}$, and $F_{Q_{H_k}}$ is sententially valid. (For more details, see van Heijenoort (1968, 1975, 1982).)

Herbrand expansion is a more general case of the infinite conjunction used by Skolem to obtain his results on $\aleph_0$-satisfiability. It allows Herbrand to begin with a formula written in Skolem (prenex) normal form and then delete all quantifiers in the normal-form formula by expansion up to k-many elements (for a k-ary domain) using conjunction for universaloid expressions and disjunction for existentialoid expressions within the formula. This method is required in Q_H because Herbrand adopted the finitism of Hilbert, for which the concept of satisfiability, especially $\aleph_0$-satisfiability, in the Löwenheim–Skolem theorem, is metamathematically unsound.

Hilbert considered ω-inconsistent systems even prior to the construction of such systems by Gödel (1931). The aim of the Hilbert program and of the technical developments in *Beweistheorie* for Hilbert was to provide a proof procedure for the consistency of full quantification theory within and by way of the axiomatic method. For Herbrand, ω-inconsistency is of itself questionable, and the concept of *ω-consistency* (indeed of *consistency*) must be modified proof-theoretically by a careful and precise definition of *satisfiability* in Q_H.

In the case of Hilbert's work on ω-inconsistency, every attempt was made to prove the consistency of arithmetic. For example, there is Hilbert's call for the axiomatization of arithmetic, together with his proof for its consistency (Hilbert, 1904) and his formalization of number theory (Hilbert, 1900). The proof-theoretic form of Gödel's 1930 completeness theorem (Gödel, *The completeness of the axioms of the functional calculus of logic* in van Heijenoort

(1967, pp. 583–591)) is given by Hilbert and Bernays (1934, 1939, vol. 2, pp. 205–253), and, in Gödel's version, makes heavy use of techniques of Löwenheim and Skolem (see van Heijenoort's introductory comments in *From Frege to Gödel*, p. 582). This attempt to provide a consistency proof for full quantification theory, particularly of arithmetic, even taking into account the completeness results of Gödel, was made at the same time by Herbrand (1931).

The Hilbert program, as developed by the Hilbert–Bernays *Beweistheorie*, seeks to provide *finitist* constructive proofs for elementary arithmetic $\mathscr{P}_0$. Under a finitist program, as taken over by Herbrand, and particularly thanks to his Fundamental Theorem, Herbrand found ω-*consistency/inconsistency* to be a metamathematically unsound, and spurious, concept. It is easy to see why. Herbrand's (1930a) objections are leveled first against Löwenheim's (1915) intuitive formulation of the notion "true in an infinite domain". Next, he argued that Hilbert's concept of completeness (*Vollständigkeit*) did not clarify the concept of *being a proof* (van Heijenoort, 1967, Burton Dreben's not H); his objection is based on the fact that consistency fails to imply provability, and that for Hilbert, completeness implies decidability but consistency does not imply provability. Thus, in his study of Hilbert's *Beweistheorie*, von Neumann (1927) concentrates his attention on the question of noncontradiction (*Widerspruchsfreiheit*). Herbrand's task was to define provability in terms of validity. He therefore explicitly took for himself the task of associating completeness with validity.

By establishing within $\mathscr{P}_0$ the conditions sufficient for formalization of $\mathscr{P}_0$ as given by Kreisel (1968, p. 322), we elaborate at the same time the relation between completeness and validity for a system S and Bew_S, the proof predicate of S:

(a) $A \rightarrow \text{Bew}_S(\pi A, A_T)$ for $A \in \hat{A}$, where $\hat{A}$ is a class of assertions of $\mathscr{P}_0$, and A_T is a formula of the formal system S, and π is a function from $\hat{A}$ into formal derivations of S;

(b) $\text{Bew}_S(f, A_T) \rightarrow A$ for $A \in \hat{A}$ and f a variable.

If these two conditions are satisfied, we established the adequacy of $\mathscr{P}_0$ and obtain finitist proofs in the system S of every $A_T \in \hat{A}$ of $\mathscr{P}_0$. Thus, every theorem A_T of $\mathscr{P}_0$ is decidable.

Perhaps the most representative feature of Q_H is its elimination of *modus ponens*. This rests on the nature of Herbrand expansions, that is, Herbrand conjunctions and disjunctions, and on the set of connectives of Q_H, with Herbrand expansion (conjunction), Herbrand disjunction, and negation being a base.[1] Gentzen's (1934) Sequent Calculus rests on the results of Herbrand, but goes further by giving an analysis of the sentential parts of the proof of validity. Also, unlike Q_H, the Gentzen theory contains a *Hauptschnitt* which is extended to the intuitionistic calculus and to modal logic. These are perhaps some of the reasons for Girard's (1982, p. 29) claim that "Herbrand's 'Théorème Fondamental', one of the milestones of mathematical logic, is also one of the few basic results of proof-theory, only challenged (and presumably overtaken) by Gentzen's Hauptsatz." (Girard then goes on to examine the current import of Herbrand's theorem for proof theory.)

Gentzen (1934, p. 409, ftn.) himself claims that Herbrand's fundamental theorem is just a special case of his own *verschärfter Hauptsatz*, insofar as Herbrand's theorem applies to a sequent whose antecedent is empty and whose succedent is a single prenex formula, while the *verschärfter Hauptsatz* applies to any prenex formulas. (The *Hauptsatz*, to remind readers, presents cut-elimination, and is analogous to Herbrand's elimination of *modus ponens* for Q_H.) However, Gentzen's cut-elmination for prenex formulas is not as general or as strong as the Herbrand elimination of modus ponens for Q_H. Statman (1977) uses the cut-elimination analogue of Herbrand's Fundamental Theorem to elucidate what can be learned from a direct proof of a theorem in a formal calculus. He takes Herbrand's theorem as a corollary to the Completeness Theorem; then using the cut-elimination technique, he proves completeness within the Kreisel–Tait equation calculus. Statman thereby helps, indirectly, to elucidate the connections between Q_H and the Gentzen Sequent Calculus.

Unlike Q_H and Gentzen's Sequent Calculus, the method of natural deduction places extreme emphasis on *modus ponens* as a method of inference, and in this regard it returns to the conditionalization introduced into the *Begriffsschrift* by Frege as the principal logical connective and the rule of detachment of the Hilbert-type axiomatic method. The central features of the method of natural deduction are the rule of conditionalization and quantifier rules. The Gentzen Sequent Calculus features rules for introduction and elimination of connectives and quantifiers, and in this has some affinity with the natural deduction method. The use of the two rules of universal generalization (UG) and existential instantiation (EI) reintroduce into $Q_=$, quantification theory with identity and prenex normal form, serious difficulties with respect to the satisfiability or even decidability of certain theorems of $Q_=$.

It was Jaśkowski (1934) who, in 1934, first utilized the idea of Łukasiewicz (1925) to found proofs on the rules of intuitive logic.[2] Jaśkowski's system uses UG and universal instantiation (UI), and is therefore considered to be insufficiently natural, since it does not provide for existential quantifiers (see Anellis (1990b)). The *N-sequenzen* (natural sequences) were found to be more intuitively acceptable, allowing existential as well as universal quantification. Thus, the Gentzen method, employing *N*-sequents, appeared to be preferable to the unnatural natural deduction of Jaśkowski, and there was a consequent rush to naturalize and save Jaśkowski's method. Ironically, Łukasiewicz proposed natural deduction as an alternative to the axiomatic systems just because he thought that those systems were unnatural, not only because they exhibited difficulties with restrictions on acceptable quantifiers, but also because they did not appear to bear any resemblance to the way the working mathematician actually obtained his results. The inference rules which were delineated by Hilbert were thought, then, to be artificial. J. Fang (1984) raises today the same objections to Gentzen's system as Jaśkowski—and for that matter, Gentzen—raised to Hilbert's axiomatic method. To these objections, he also adds (1984, p. 14) the complaint that Gentzen's system is unnatural because it is too complicated and contains an over-abundance of inference rules; he even claims that Gentzen himself was "perfectly aware of the 'unnatural' aspect in some artificial rules of inference," citing in particular Gentzen's rules (3a) and (3b) (i.e., the laws of addition: $P \to (P \lor Q)$ and $Q \to (P \lor Q)$ (Gentzen, 1934, p. 186)). But then Fang is among those who would prefer a return to classical Aristotelean syllogistic.

Despite the work of Quine (1955), Prawitz (1965), and others, UG and EI persist as sore spots for natural deduction.[3] We are led to suppose that, among alternative systems of natural deduction, the N-sequents of Gentzen are preferable to other, particularly Copi-type, deductions. (A history of the vicissitudes and problems encountered in attempts to devise quantification rules for natural deduction is given by Anellis (1990b).)

The problems encountered in formulating quantifier rules, particularly the problems arising from UG and EI in Copi-style natural deduction systems, returns us to the question of satisfiability for quantified formulas involving these two troublesome and elusive rules. We can prove, for example, that some formula F' of $Q_=$ is k-satisfiable for any finite k but is not $(k-1)$-valid; or that some formula F'' of $Q_=$ is k-satisfiable for any finite k but is not $\aleph_0$-satisfiable. For $Q_=$, under UG and EI, we face the Law of Lesser Universes. According to this law, for $0 < \alpha \le \beta$ cardinals, if a formula F is satisfiable in a universe of cardinality α, it is also β-satisfiable; and if F is β-valid, it is also α-valid. Now the Law of Lesser Universes applies only to those quantification theories in which formulas occur in disjunctive normal form, and in which there is neither identity nor a universal quantifier. However, for $Q_=$, satisfiability and validity depend largely on the cardinality of the universe of discourse. The questions raised by natural deduction have been rendered less sharp by the introduction by Prawitz (1968) into the natural deduction systems of non-Gentzen type, and especially into higher-order predicate calculus, of an analogue of Gentzen's *Hauptsatz* on cut-elimination (*Schnitt*-elimination).

In his important work, Takeuti (1975) has done much with the Gentzen sequential method, and it is perhaps to be preferred to the Hilbert inferential method. Schütte (1960, 1977) takes Hilbert's method as a starting point, and is nearly indistinguishable from Hilbertian *Beweistheorie*. Takeuti's work in proof theory may even be preferable to that of Schütte, and indeed can be seen as a third alternative to both the Hilbert and Gentzen methods, in fact, combining the best features of both. (Feferman (1977, 1979) has reviewed both Takeuti (1975) and Schütte (1977).)

Quantification theory, starting from the definition of the Hilbert program and the construction of Herbrand quantification, through the Gödel incompleteness results to the Gentzen sequences and natural deduction, presents a "family of formal systems," a development of quantification theory, in the attempt to clarify the concepts of *proof*, *satisfiability*, and *validity*. Over the years, some progress may even be presumed to have been made. The two works by Takeuti and Schütte give important new results, some of which are pointed out by Feferman. In spite of the differences in style between them, with Schütte's *Beweistheorie* strongly influenced by the Hilbert-type system and Takeuti's *Proof Theory* belonging strictly to the Gentzen-sequent system, they tend to complement and complete one another. It was the problems inherent in the Löwenheim–Skolem theorem concept of $\aleph_0$-*satisfiability* that served as a prime stimulus for work in proof theory and in the development of technical results in quantification theory, including Herbrand's Q_H, Gentzen's Sequent Calculus, and the other systems of the Natural Deduction method whose principal contributors include Jaśkowski, Prawitz, and Quine. The reader may also be reminded that the Semantic Tableaux of Beth are a semanticization of the Gentzen Sequence Calculus, and that the Truth Tree method introduced by Jeffrey (1967) is a simplification of Beth Tableaux through the intermediacy of Hugues Leblanc's application of semantic tableaux to natural deduction. We need not, therefore, list these technical innovations as separate members of the family of formal systems, but must note that Jeffrey (1967, p. ix) himself traces the Tree Method back even to Herbrand. For a complete history of the tree method, see Anellis (1990a).

Interestingly, and as I had noted earlier, van Heijenoort (1977, p. 183, 185) reminds us that Löwenheim had no notion of *proof*, just of *satisfiability*, and that Skolem had been seeking a proof procedure for completeness of Hilbert-type quantification theory. It is certainly clear that the weakness of Hilbert-type quantification theory became evident in and because of the Löwenheim–Skolem theorem. It was precisely this weakness that became fully explicit in Herbrand expansion as an enlargement of the Löwenheim–Skolem infinite conjunction presented by Skolem in his normal-form translations of Hilbert's quantified formulas, and which, I claim, thereby led to important work in Hilbert's *Beweistheorie* specifically, and in proof theory generally, and to the construction of alternative theories of quantification, to systems other than axiomatic theories.

Notes

1. For an axiom system (or for any theory of quantification), the [smallest] set of primitive logical connectives, in terms of which other logical connectives come to be defined, is called a *base*. For Frege's system, negation and implication constitute the base; for the *Principia Mathematica* system, negation and disjunction constitute the base.

In Q_H, a *base* can be understood as the set of connectives in virtue of which existentialoidally quantified formulas can be expanded (Herbrand disjunction) and by which universaloidally quantified formulas can be expanded (Herbrand conjunction), together with negation.

2. See Anellis (1990b, p. 5). Jaśkowski (1934, p. 5) is more specific and refers to Łukasiewicz's 1926 seminar, in which some of the ideas were first worked out, and to their presentation at the First Polish Mathematical Congress in Lwów in 1927 and

subsequent publication in the Congress Proceedings (*Ksiega pamiatkowa pierwszego polskiego zjazdu matematycznego*, Kraków, 1929).

3. I do not number Copi among those whose work should be considered. The notorious history of his revisions for EI and UG in successive editions of his *Symbolic Logic* has been chronicled by Anellis (1990b).

References

Anellis, I.H. (1979), The Löwenheim–Skolem theorem, theories of quantification, and Beweistheorie (Abstract), *Notices Amer. Math. Soc.* **26**, A-22.

Anellis, I.H. (1990a), From semantic tableaux to Smullyan trees: the history of the falsifiability tree method, *Modern Logic* **1**, 36–69.

Anellis, I.H. (1990b), Forty years of unnatural natural deduction and quantification: a history of first-order systems of natural deduction, from Gentzen to Copi, preprint.

Fang, J. (1984), The most unnatural "natural deduction," *Contemp. Philos.* **10**, no. 3, 13–15.

Feferman, S. (1977), Review of *Proof Theory* by Gaisi Takeuti, *Bull. Amer. Math. Soc.* (1), **83**, 351–361.

Feferman, S. (1979), Review of *Proof Theory* by Kurt Schütte, *Bull. Amer. Math. Soc.* (2), **1**, 224–228.

Frege, G. (1879), *Begriffsschrift, eine der arithmetischen nachgebildete Formelsprache des reinen Denkens*, L. Nebert, Halle.

Frege, G. (1883), *Über den Zweck der Begriffsschrift*, Sitzungsberichte der Jenaischen Gesellschaft für Medicin und Naturwissenschaft fur des Jahr 1882, pp. 1–10. (English translation by V.H. Dudman in *Austral. J. Philos.* **46** (1968), 89–97.)

Gentzen, G. (1934), Untersuchungen über das logische Schliessen, *Math. Z.* **39**, 176–210, 405–431.

Girard, J.-Y. (1982), Herbrand's theorem and proof theory, in J. Stern (ed.), *Proc. Herbrand Symposium, Logic Colloquium '81*, North-Holland, Amsterdam, pp. 29–38.

Gödel, K. (1931), Über formal unentscheidbare Sätze der Principia Mathematica und verwändter Systeme. I. *Monatsh. Math. Phys.* **38**, 173–198. (English translation in van Heijenoort (1976a), pp. 581–591.)

Goldfarb, W.D. (ed.) (1971), *Jacques Herbrand. Logical Writings*, Harvard University Press, Cambridge, MA.

Goldfarb, W.D. (1979), Logic in the twenties: the nature of the quantifier, *J. Symbolic Logic* **44**, 351–368.

Herbrand, J. (1930a), *Recherches sur la théorie de la démonstration*, Prace Towarzystwa Naukowege Warszawskiego, Wydział III, 33. (English translation in Goldfarb (1979), pp. 46–202; translation of Chapter 5 in van Heijenoort (1976a), pp. 529–581.)

Herbrand, J. (1930b), Les bases de la logique hilbertienne. *Revue de Métaphysique et de Morale* **37**, 243–255. (English translation in Goldfarb (1979), pp. 203–214.)

Herbrand, J. (1931), Sur la non-contradiction de l'arithmétique, *J. Reine Angew. Math.* **166**, 1–8. (English translation in Goldfarb (1979), pp. 282–298.)

Hilbert, D. (1900), Über den Zahlbegriff. *Jahresb. Deutsch. Math.-Verein.* **8**, 180–194.

Hilbert, D. *Über die Grundlagen der Logik und Mathematik*, Verhandlungen des dritten internationalen Mathematiker-Kongress in Heidelberg (1904), pp. 174–185. (English translation in van Heijenoort (1976a), pp. 129–138.)

Hilbert, D. (1918), Axiomatisches Denken, *Math. Ann.* **78**, 405–415. (Reprinted in *David Hilbert, Gesammelte Abhandlungen*, III, 2nd ed., Springer-Verlag, Berlin, 1970, pp. 146–156.)

Hilbert, D. and Bernays, P. (1934, 1939), *Grundlagen der Mathematik*, 2 vols., Springer-Verlag, Heidelberg.

Jaśkowski, S. (1934), On the rules of supposition in formal logic, *Studia Logica* 1, 5–32.

Jeffrey, R.C. (1967), *Formal Logic: Its Scope and Limits*, McGraw-Hill, New York.

Kreisel, G. (1959), *Hilbert's programme, Logica: Studia Paul Bernays dedicata*, editions du Griffon, Neuchatel, pp. 142–167.

Kreisel, G. (1968), A survey of proof theory, I, *J. Symbolic Logic* **33**, 321–388.

Kreisel, G. (1971), A survey of proof theory, II, in J.E. Fenstad (ed.), *Proc. 2nd Scandin. Logic Symposium*, North-Holland, Amsterdam, pp. 109–170.

Kreisel, G. (1976), What have we learnt from Hilbert's second problem? in F. Browder (ed.), Proc. Symposium Pure Math., 28, *Mathematical Developments Arising From Hilbert Problems*, American Mathematical Society, Providence, RI, pp. 93–130.

Löwenheim, L. (1915), Über Möglichkeiten im Relativkalkül, *Math. Ann.* **76**, 447–470. (English translation in van Heijenoort (1976a), pp. 323–251.)

Łukasiewicz, J. (1925), O pewnym sposobie pojmowania (On a certain way of conceiving the theory of deduction), *Przeglad Filozoficzny* **28**, 134–136.

Minc, G.E. [Mints] (1975), Teorija dokazatel'sty (arifmetika i analiz), *Itogi Nauki i Tehniki. Algebra, Topologija, i Geometrija* 13, 5–49. (English translation: Theory of proofs (arithmetic and analysis), in *J. Soviet Math.* 7, no. 4, 501–531.)

Prawitz, D. (1965), *Natural Deduction: A Proof-Theoretical Study*, Almqvist & Wiksell, Stockholm.

Prawitz, D. (1968), Hauptsatz for higher-order logic, *J. Symbolic Logic* **33**, 452–457.

Prawitz, D. (1971), Ideas and results in proof theory, in J.E. Fenstad (ed.), *Proc. 2nd Scand. Logic Symposium*, North-Holland, Amsterdam, pp. 235–307.

Quine, W.V. (1955), A proof procedure for quantification theory, *J. Symbolic Logic* **20**, 141–149 (Reprinted in W.V. Quine, *Selected Logic Papers*, Random House, New York, 1966, pp. 196–204.)

Schütte, K. (1960), *Beweistheorie*, Springer-Verlag, Berlin.

Schütte, K. (1977), *Proof Theory*, Springer-Verlag, Berlin, revised edition of Schütte (1960), J.N. Crossley (trans.).

Skolem, T. (1920), *Logisch-kombinatorische Untersuchungen über die Erfühlbarkeit und Beweisbarkeit mathematischen Sätze nebst einem Theoreme über dichte Menge*, Videnkasselskapets skrifter, I, Matematik-naturvidenskabelig klasse 4. (Reprinted in J.E. Fenstad (ed.), *Selected Works in Logic by Th. Skolem*, Universitetsforlaget, Oslo, 1970, Sect. 1, pp. 103–115; English translation in van Heijenoort (1976a), pp. 254–263.)

Skolem, T. (1923), *Einige Bemerkung zur axiomatischen Begründung der Mengenlehre*, Matematikerkongressen i Helsingfors den 4–7 Juli, 1922, Den femte Skandinaviska matematikerkongressen, Redogörelse, Akademiska Bokhandeln, Helsinki (Helsingfors), pp. 217–232. (Reprinted in J.E. Fenstad (ed.), *Selected Works in Logic by Th. Skolem*, Universitetsforlaget, Oslo, 1970, pp. 137–152; English translation in van Heijenoort (1976a), pp. 291–301.)

Statman, R. (1977), Herbrand's theorem and Gentzen's notion of a direct proof, in

J. Barwise (ed.), *Handbook of Mathematical Logic*, North-Holland, Amsterdam, pp. 897–912.

Takeuti, G. (1975), *Proof Theory*, North-Holland/American Elsevier, Amsterdam, New York.

van Heijenoort, J. (ed.) (1967), *From Frege to Gödel: A Source Book in Mathematical Logic, 1879–1931*, Harvard University Press, Cambridge, MA.

van Heijenoort, J. (1968), Préface, in J. van Heijenoort (ed.), *Jacques Herbrand, Ecrits Logiques*, Presses Universitaires de France, Paris. pp. 1–12.

van Heijenoort, J. (1975), *Herbrand*, Unpublished MS.

van Heijenoort, J. (1976), *El desarrollo de la teoría de la cuantificación*, Universidad nacional autónoma de México, Mexico City.

van Heijenoort, J. (1977), Set-theoretic semantics, in R. Grandy and M. Hyland (eds.), *Logic Colloquium 1976*, North-Holland, Amsterdam, pp. 183–190.

van Heijenoort, J. (1982), L'oeuvre de Jacques Herbrand et son contexte historique, in J. Stern (ed.), *Proc. Herbrand Symposium, Logic Colloquium '81*, North-Holland, Amsterdam, pp. 57–85.

von Neumann, J. (1927), Zur Hilbertschen Beweistheorie, *Math. Z.* **26**, 1–46.

Whitehead, A.N. and Russell, B. (1910–1913), *Principia Mathematica*, 3 vols., Cambridge University Press, Cambridge, UK.

The Reception of Gödel's Incompleteness Theorems

John W. Dawson, Jr.

"Die Arbeit über formal unentscheidbare Sätze wurde wie ein Erdbeben empfunden; insbesondere auch von Carnap."—(Popper, 1980).

"Kurt Gödel's achievement in modern logic ... is a landmark which will remain visible far in space and time."—John von Neumann[1]

It is natural to invoke geological metaphors to describe the impact and the lasting significance of Gödel's incompleteness theorems. Indeed, how better to convey the impact of those results—whose effect on Hilbert's program was so devastating and whose philosophical reverberations have yet to subside—than to speak of tremors and shock waves? The image of shaken foundations is irresistible.

Yet to adopt such seismic imagery is to suggest that the aftermath of intellectual upheaval is comparable to that of geological cataclysm: a period of slow rebuilding, preceded initially by widespread confusion and despair, or perhaps determined resistance; and though we might expect Gödel's discoveries to have provoked just such reactions, according to most commentators they did not. Thus van Heijenoort states "although [Gödel's paper] caused some momentary surprise, its results were soon widely accepted" (van Heijenoort, 1967, p. 594). Similarly, Kreisel has averred that "expected objections never materialized" (Kreisel, 1979, p. 13), and Kleene, speaking of the second incompleteness theorem (whose proof was only sketched in Gödel's (1931)), has even claimed "it seems no one doubted it" (Kleene, 1976, p. 767).

If these accounts are correct, one of the most profound discoveries in the history of logic and mathematics was assimilated promptly and almost without objection by Gödel's contemporaries—a circumstance so remarkable that it demands to be accounted for. The received explanation seems to be that Gödel, sensitive to the philosophical climate of opinion and anticipating objections to his work, presented his results with such clarity and rigor as to

PSA 1984, Volume 2,
Copyright © 1985 by the Philosophy of Science Association.
Reprinted with permission.

render them incontestable, even at a time of fervid debate among competing mathematical philosophies. The sheer force of Gödel's logic, as it were, swept away opposition so effectively that Gödel abandoned his stated intention of publishing a detailed proof of the second theorem (Gödel, 1931, p. 198).

On the last point there can be no dispute, as Gödel stated explicitly to van Heijenoort that "the prompt acceptance of his results was one of the reasons that made him change his plan" (van Heijenoort, 1967, ftn. 68a, p. 616). We may question, however, to what extent Gödel's subjective impression reflected objective circumstances. We must also recognize the hazard in assessing the cogency of Gödel's arguments from our own perspective. To be sure, the exposition in Gödel (1931) *now* seems clear and compelling; the proofs strike us as detailed but not intricate. But did it seem so at the time? After all, arithmetization of syntax was then a novel device, and logicians were not then so accustomed to the necessity for making precise distinctions between object- and metalanguage. Indeed, J. Barkley Rosser (who himself contributed to the improvement of Gödel's results) has observed that only *after* Gödel's paper appeared did logicians realize how careful they had to be (cf. Grattan-Guinness, 1981, ftn. 3, p. 499); precisely *because* of Gödel's results we no longer share the formalists' naive optimism, and so we are likely to be more receptive to Gödel's ideas.

Our faith in the efficacy of logic as a tool for overcoming intellectual resistance should also be tempered by consideration of the reception accorded other "paradoxical" results. For example, the Löwenheim–Skolem theorem, first enunciated by Löwenheim in 1915 and established with greater precision and broader generality by Skolem in a series of papers from 1920 to 1929, is certainly less profound than Gödel's discovery (with which, however, it is still sometimes confused)—yet it generated widespread misunderstanding and bewilderment, even as late as December 1938. (See, in particular, the discussion in Gonseth (1941, pp. 47–52).)

In what follows, I shall examine the reaction to Gödel's theorems in some detail, with the aim of showing that there *were* doubters and critics, as well as defenders and rival claimants to priority. (Of course, there were also some who *accepted* Gödel's results without fully *understanding* them.)

1. 1930: Announcement at Königsberg

Elsewhere (Dawson, 1984) I have described in detail the circumstances surrounding Gödel's first public announcement of his incompleteness discovery. In summary, the event occurred during a discussion on the foundations of mathematics that took place in Königsberg, 7 September 1930, as one of the final sessions of the Second Conference on Epistemology of the Exact Sciences organized by the *Gesellschaft für empirische Philosophie*. At the time, Gödel was virtually unknown outside Vienna; he had come to the conference to deliver a 20-minute talk on the results of his dissertation, completed the year before and just then about to appear in print. In that work, Gödel had established a result of prime importance for the advancement of Hilbert's program: the completeness of first-order logic (or, as it was then called, the restricted functional calculus); so it could hardly have been expected that the day after his talk Gödel would suddenly undermine that program by asserting the existence of formally undecidable propositions. As Quine has remarked, "[although] completeness was expected, an actual proof of completeness was less expected, and a notable accomplishment. It came as a welcome reassurance.... On the other hand the incompletability of elementary number theory came as an upset of firm preconceptions" (Quine, 1978, p. 81).

To judge from the (edited) transcript of the discussion published in *Erkenntnis* (Hahn et al., 1931), none of the other participants at Königsberg had the slightest inkling of what Gödel was about to say, and the announcement itself was so casual that one suspects that some of them may not have realized just what he *did* say. In particular, the transcript gives no indication of any discussion of Gödel's remarks, and there is no mention of Gödel at all in Reichenbach's post-conference survey of the meeting (published in *Die Naturwissenschaften* 18: 1093–1094). Yet two, at least, among those present *should* have had foreknowledge of Gödel's results: Hans Hahn and Rudolf Carnap.

Hahn had been Gödel's dissertation advisor. He chaired the discussion at Königsberg, and it was presumably he who invited Gödel to take part. Of course, Gödel might not have confided his new discovery to him—indeed, Wang has stated that "Gödel completed his dissertation before showing it to Hahn" (Wang, 1981, ftn. 2, p. 654)—but in introductory remarks to the dissertation that were *deleted* from the published version (at whose behest we do not know), Gödel had explicitly raised the possibility of incompleteness (without claiming to have demonstrated it). Perhaps Hahn just didn't take the possibility seriously.

In any case, Gödel *did* confide his discovery to Carnap prior to the discussion at Königsberg, as we know from *Aufzeichnungen* in Carnap's *Nachlass*. Specifically, on 26 August 1930, Gödel met Carnap, Feigl, and Waismann at the Cafe Reichsrat in Vienna, where they discussed their travel plans to Königsberg. Afterward, according to Carnap's entry for that date, the discussion turned to "Gödels Entdeckung: Unvollständigkeit des Systems der PM; Schwierigkeit des Widerspruchsfreiheitbeweises."[2] Three days later another meeting took place at the same cafe. On that occasion, Carnap noted "Zuerst [before the arrival of Feigl and Waismann] erzählt mir *Gödel* von seiner Entdeckungen." Why then at Königsberg did Carnap persist in advocating consistency as a criterion of adequacy for formal theories? That he might have done so just to provide an opening for Gödel seems hardly credible. It seems much more likely that he simply failed to *understand* Gödel's ideas. (As it happens, a subsequent note by Carnap dated 7 February 1931, after the appearance of Gödel's paper, provides confirmation: "Gödel hier. Über seine Arbeit, ich sage, dass sie doch schwer verständlich ist.") Later, of course,

Carnap was among those who helped publicize Gödel's work; but Popper's remark quoted at the head of this article seems to be an accurate characterization of Carnap's initial reaction.

One of the discussion participants who did immediately appreciate the significance of Gödel's remarks was John von Neumann: After the session he drew Gödel aside and pressed him for further details. Soon thereafter he returned to Berlin, and on 20 November he wrote Gödel to announce his discovery of a remarkable [*bemerkenswert*] corollary to Gödel's results: the unprovability of consistency. In the meantime, however, Gödel had himself discovered his second theorem and had incorporated it into the text of his paper; the finished article was received by the editors of *Monatshefte* 17 November.[3]

2. 1931: Publication and Confrontation

In January 1931 Gödel's paper was published. But even before then, word of its contents had begun to spread. So, for example, on 24 December 1930, Paul Bernays wrote Gödel to request a copy of the galleys of (1931), which Courant and Schur had told him contained "bedeutsamen und überraschenden Ergebnissen." Gödel responded immediately, and on 18 January 1931, Bernays acknowledged his receipt of the galleys in a 16-page letter in which he described Gödel's results as "ein erheblicher Schritt vorwärts in der Erforschung der Grundlagen probleme."

The Gödel–Bernays correspondence is of special interest (in the absence of more direct evidence[4]) for the light it sheds on Hilbert's reaction. In the same letter of 18 January, Bernays discussed Hilbert's "recent extension of the usual domain of number theory"—his introduction of the ω-rule—and von Neumann's belief that every finitary method of proof could be formalized in Gödel's system P. Bernays himself saw Gödel's theorem as establishing a disjunction: either von Neumann was right, and no finitary consistency proof was possible for the systems Gödel considered, or else some finitary means of proof were not formalizable in P—a possibility Gödel had expressly noted in his paper. In any case, Bernays felt, one was impelled [*hingedrängt*] to weaken Gödel's assumption that the class of axioms and the rules of inference be (primitively) recursively definable. He suggested that the system obtained by adjoining the ω-rule would escape incompleteness yet might be proved consistent by finitary means. But (according to Carnap's *Aufzeichnung* of 21 May 1931) Gödel felt that Hilbert's program would be compromised by acceptance of the ω-rule.

Somewhat later Gödel sent Bernays an offprint of the incompleteness paper, enclosing a copy for Hilbert as well. In his acknowledgement of 20 April, Bernays confessed his inability to see why a truth predicate could not be formally defined in number theory—he went so far as to propose a candidate for such a definition—and why Ackermann's consistency proof (which he had accepted as correct) could not be formalized there as well. By 3 May, when he wrote Gödel once more, Bernays had recognized his errors, but the correspondence remains of interest, not only because it exposes Bernays' difficulties in assimilating the consequences of Gödel's theorems, but because it furnishes independent evidence of Gödel's awareness of the formal undefinability of the notion of truth—a fact nowhere mentioned in (1931).[5]

Gödel's formal methods, as employed in the body of (1931), thus seem to have served their purpose in securing the acceptance of his results by three of the leading formalists. But at the same time, even among those who appreciated the value of formalization, Gödel's precise specification of the system P raised doubts as to the *generality* of his conclusions.[6] On the other hand, those opposed to formal systems could point to Gödel's results as reason for dismissing such systems altogether.

Partly to obviate such objections, Gödel soon extended his results to a wider class of systems (in (1930/31) and (1934)), and in the introduction to (1931) he also gave *informal* proofs of his results based on the soundness of the underlying axiom systems rather than on their formal consistency properties. Undoubtedly, he hoped this informal précis would help his readers to cope with the formal detail to follow; but all too many read no further. Ironically, because of the misinterpretations it subsequently spawned, the introduction to (1931) has been called that paper's "one 'mistake'" (Helmer, 1937).

During 1931, Gödel spoke on his incompleteness results on at least three occasions: at a meeting of the Schlick circle (15 January), in Karl Menger's mathematics colloquium (22 January), and, most importantly, at the annual meeting of the *Deutsche Mathematiker-Vereinigung* in Bad Elster (15 September), where, in Ernst Zermelo, he encountered one of his harshest critics.

At issue between the two men were profound differences in philosophy and methodology. In his own talk at Bad Elster (abstracted in Zermelo (1932)), Zermelo lashed out against "Skolemism, the doctrine that *every* mathematical theory, even set theory, is realizable in a countable model"—a doctrine that Zermelo regarded as an embodiment of Richard's antinomy. For Zermelo, quantifiers were infinitary conjunctions or disjunctions of unrestricted cardinality; and since the truth values of compound statements were therefore determined by transfinite induction on the basis of the truth values assigned to the *Grundrelationen,* Zermelo argued that *this determination itself* constituted the proof or refutation of each proposition. There were no "undecidable" propositions, simply because Zermelo's infinitary logic had no syntactic component. Consequently, Zermelo dismissed Gödel's "attempt" to exhibit undecidable propositions, saying that Gödel "applied the 'finitistic' restriction only to the 'provable' statements" of his system, "not to *all* statements belonging to it,"[7] so that Gödel's result, like the Löwenheim–Skolem theorem, depended on (unwarranted) cardinality restrictions. It said nothing about the existence of "absolutely unsolvable problems in mathematics."

In his published remarks, Zermelo did not fault the correctness of Gödel's argument; he merely took it as evidence of the untenability of the "'finitistic' restriction." But on 21 September, immediately following the conference, Zermelo wrote privately to Gödel to inform him of "an essential gap" [*eine wesentliche Lücke*] in his argument. (See Dawson (1985) for the full text of this letter.) Indeed, Zermelo argued, simply by omitting the proof predicate from Gödel's construction one would obtain a formal sentence asserting its own falsity, yielding thereby "a *contradiction* analogous to Russel[1]'s antinomy." Zermelo's letter prompted a further exchange of letters between the two men (published in Grattan-Guinness (1979)), with Gödel patiently explaining the workings of his proof, pointing out the impossibility of defining truth combinatorially within his system, and emphasizing that the introductory pages of his paper (to which Zermelo had referred) did not pretend to precision, as did the detailed considerations later on. In reply, Zermelo thanked Gödel for his letter, from which, he said, he had gained a better understanding of what Gödel meant to say; but in his published report he still failed utterly to appreciate Gödel's distinctions between syntax and semantics.

3. Recognition, and Challenges to Priority

Despite (or perhaps because of) Zermelo's reputation as a polemicist, his crusade against "Skolemism" won few adherents, and his criticisms of Gödel's work seem to have been disregarded. In the spring of 1932 Karl Menger became the first to expound the incompleteness theorem to a popular audience, in his lecture *Die neue Logik* (published by Franz Deuticke in 1933 as one of "fünf Wiener Vorträge" in the booklet *Krise und Neuaufbau in den exakten Wissenschaften*; see Menger (1978) for an English translation). The following June Gödel submitted (1931) to the University of Vienna as his *Habilitationsschrift*, and in January 1933 he accepted an invitation to spend the academic year 1933–34 at the newly established Institute for Advanced Study in Princeton.

On 11 March 1933, Gödel's *Dozentur* was granted. That same day Paul Finsler in Zürich addressed a letter to Gödel requesting a copy of (1931). He was interested in Gödel's work, he said, because it seemed to be closely related to earlier work of his own (Finsler, 1926). He had already glanced fleetingly at Gödel's paper, and he acknowledged that Gödel had employed "a narrower and therefore sharper formalism." It was "of course of value," he conceded, "actually to carry out [such] ideas in a specific formalism," but he had refrained from doing so because he felt that he had already established the result in a way that "went further in its application to Hilbert's program."

Gödel recognized the challenge for what it was, and in his reply of 25 March he characterized Finsler's system as "not really defined at all" [*überhaupt nicht definiert*], declaring that Finsler's ideas could not be carried out in a genuinely formal system, since the antidiagonal sequence defined by Finsler (on which his undecidable proposition was based) would never be representable within that same system. Finsler retorted angrily that it was not necessary for a system to be "sharply" defined in order to make statements about it; it was enough that "one be able to accept it as given and to recognize a few of its properties." He could, he said, "with greater justice" object to Gödel's proof on the grounds that Gödel had not shown the Peano axioms that he employed to be consistent (Gödel's second theorem notwithstanding!). He saw that the truth of Gödel's undecidable sentence could only be established metamathematically and so concluded that there was "no difference in principle" between his and Gödel's examples.

Van Heijenoort (1967, pp. 438–440) has analyzed Finsler's paper in detail, concluding that it remains a sketch whose "affinity [with Gödel's paper] should not be exaggerated." In effect, whereas "Gödel put the notion of formal system at the very center of his investigations," Finsler rejected such systems as artificially restrictive (prompting Alonzo Church to remark dryly that such "restricted" notions have at least the merit of being precisely communicable from one person to another") (Church, 1946). Instead, especially in his (1944), Finsler attempted to show the consistency of the assertion that were no "absolutely undecidable" propositions.

In contrast, Emil Post directed his efforts toward showing that there *were* absolutely unsolvable problems in mathematics.[8] Early on, nearly a decade before Gödel, Post realized that his methods could be applied to yield a statement undecidable within *Principia* whose truth could nevertheless be established by metamathematical considerations. Hence, Post concluded, "mathematical proof was [an] essentially creative [activity]" whose proper elucidation would require an analysis of "all finite processes of the human mind"; and since the implications of such an analysis could be expected to extend far beyond the incompleteness of *Principia*, Post saw no reason to pursue the latter.

Concern for the question of *absolute* undecidability thus led Finsler, Post, and Zermelo, in varying directions, away from consideration of particular formal systems. Unlike Finsler and Zermelo, however, Post expressed "the greatest admiration" for Gödel's work, and he never sought to diminish Gödel's achievement. Indeed, Post acknowledged to Gödel that nothing he had done "could have replaced the splendid actuality of your proof," and that "after all it is not ideas but the execution of ideas that constitute[s] ... greatness." Post's (1965) "Account of an Anticipation" was not submitted until 1941 and (after being rejected) only finally appeared in print in 1965, 11 years after Post's death.[9]

It is worth noting Gödel's own opinion of the notion of absolute undecidability, as expressed in the unsent letter draft cited in note 5:

> As for work done earlier about the question of formal decidability of mathematical propositions, I know only a paper by Finsler .... However, Finsler omits exactly the main point which makes a proof possible, namely restriction to some well-defined formal system in which the proposition is undecidable. For he had the nonsensical aim of proving formal undecidability in an absolute sense. This leads to

[his] nonsensical definition [of a system of signs and of formal proofs therin] ...
and to the flagrant inconsistency that he decides the "formally undecidable proposition by an argument ... which, *according to his own definition ... is a formal proof*.
If Finsler had confined himself to some well-defined formal system *S*, his proof ...
could [with the hindsight of Gödel's own methods] be made correct and applicable
to any formal system. I myself did not know his paper when I wrote mine, and other
mathematicians or logicians probably disregarded it because it contains the obvious
nonsense just mentioned. (Gödel to Yossef Balas, 27 May 1970).

Especially as applied to Finsler's later work, the epithet "obvious nonsense"
is well deserved—Finsler (1944) in particular is an almost pathological
example of the confusion that can arise from failure to distinguish between
use and mention—and Gödel was undoubtedly sensitive to any rival claim
to priority for his greatest discovery. Nevertheless, the passage quoted above
is uncharacteristically harsh, and insofar as it ridicules the idea of "proving
formal undecidability in an absolute sense" it seems to ignore Gödel's own
remarks in 1946 before the Princeton Bicentennial Conference, in which he
suggested that, despite the incompleteness theorems, "closer examination
shows that [such negative] results do not make a definition of the absolute
notions concerned impossible under all circumstances." In particular, he
noted that "by a kind of miracle" there is an absolute definition of the concept
of computability, even though "it is merely a special kind of demonstrability
or decidability"; and in fact, the incompleteness theorems may be subsumed
as corollaries to the existence of algorithmically unsolvable problems. Finsler,
of course, had no such thing in mind in 1933, nor did Gödel. But what of Post?
Had he not been hampered by manic-depression, might he have pre-empted
Gödel's results? In any case, Gödel contributed far less than Post to the
development of recursion theory, even though he gave the first definition of
the notion of general recursive function (in (1934), following a suggestion of
Herbrand). Until Turing's (1936/37) work, Gödel resisted accepting Church's
thesis (see Davis (1982) for a detailed account), and we may well wonder
whether Gödel's focus on specific formalisms did not tend to blind *him* to the
larger question of algorithmic undecidability. Gödel repeatedly stressed the
importance of his philosophical outlook to the success of his mathematical
endeavors; perhaps it may also have been responsible for an occasional
oversight. (See Kleene (1981a, b) for recent accounts of the origins of recursive
function theory and Feferman (1985) for another view of Gödel's role as a
bystander.)

4. Assimilation, and Later Criticism

After his I.A.S. lectures in the winter and spring of 1934, Gödel returned to
Vienna and turned his attention to set theory. By the fall of 1935, when he
returned briefly to the I.A.S., he had already succeeded in proving the relative
consistency of the axiom of choice. In the meantime, however, he had also

once had to enter a sanatorium for treatment of depression. Back in America, he suffered a relapse that forced him to return to Austria just two months after his arrival. He re-entered a sanatorium and did not resume teaching at the University of Vienna until the spring of 1937.

During this period of incapacitation, Gödel's incompleteness results were improved by Rosser (1936) (who weakened the hypothesis of ω-consistency to that of simple consistency) and extended, as already noted, through the development of recursion theory by Church, Kleene, and Turing. By 1936, then, the incompleteness theorems would seem to have taken their place within the corpus of firmly established mathematical facts.

In that year, however, the correctness of Gödel's conclusions was challenged in print by Charles Perelman (1936), who asserted that Gödel had in fact discovered an *antinomy*. According to Perelman, if the soundness of the underlying axiom system were assumed (as Gödel had done in his informal introductory remarks), Gödel's methods could be employed to prove two false equivalences, namely

$$\mathrm{Dem}(\sim_q Fq) \equiv \mathrm{Dem}(_q Fq)$$

and

$$\mathrm{Dem}(_q Fq) \equiv \sim \mathrm{Dem}(_q Fq),$$

where "$_q Fq$" denotes Gödel's undecidable sentence and "Dem" denotes Gödel's provability predicate (called "Bew" in Gödel (1931)). The second statement is obviously a contradiction, and Perelman proposed to exorcise it by the radical expedient of rejecting the admissibility of the set of Gödel numbers of unprovable sentences.

Perelman's equivalences display a rather obvious conflation of object- and metalanguage, and, as Kleene noted in his review of Perelman (1936), if expressed without abuse of notation, "the first of them is not false, and the second is not deducible" (Kleene, 1937a)—indeed, properly formulated, the first equivalence is just the statement that "$_q Fq$" *is* formally undecidable. It would thus be tempting to dismiss Perelman as a crank, were it not that his arguments were apparently taken quite seriously by many within the mathematical community—so much so, in fact, that two individuals, Kurt Grelling and Olaf Helmer, felt obliged to come to Gödel's defense.

Grelling was among those who had traveled with Gödel to Königsberg in 1930. On 2 February, 1937, he wrote Gödel to advise him of Hempel's report that "angesehene Mathematiker" in Brussels and Paris were being "taken in" [*hereingefallen*] by Perelman's arguments. If Gödel were not already planning to publish a rebuttal, Grelling requested permission to do so on his behalf. Gödel did not in fact enter into the controversy, and Grelling's (1937/38) article appeared at about the same time as Helmer's (1937).

Grelling began his reply to Perelman by giving an informal outline of Gödel's proof, drawing a careful distinction between arithmetical statements and their metamathematical counterparts. Through the process of arithmetization, he explained, a metamathematical statement Q' was associated to Gödel's arithmetical statement Q, in such a way that to any formal proof of Q there would correspond a metamathematical proof of Q'; but, he asserted, a *metamathematical* proof or refutation of Q' would lead directly to a *metamathematical* contradiction ("Sowohl ein Beweis von Q' als auch eine Widerlegung würde unmittelbar auf einen Widerspruch führen.") (Grelling, 1937/38, p. 301), and so Q itself must be formally undecidable. Grelling went on to analyze Perelman's formal arguments, especially his claims that

$$(n). \mathrm{Dem}(_nFn). \supset . \mathrm{Dem}(\mathrm{Dem}(_nFn)) \qquad (1)$$

and

$$(n). \sim \mathrm{Dem}(_nFn). \supset . \mathrm{Dem}(\sim \mathrm{Dem}(_nFn)) \qquad (2)$$

were provable in Gödel's system. Without commenting on the abuse of notation involved (whereby "Dem" should correspond to Gödel's *meta-mathematical* predicate "Bew," not to the Gödel number of its arithmetical counterpart), he (correctly) accepted (1) as demonstrable, while rejecting (2).

Helmer, on the other hand, stressed Perelman's notational confusion, pointing out that "'Dem' is a predicate applicable to numbers only," so that "if formula (1) is to be significant, the expression '$_nFn$' must denote a number and not a sentence." One might therefore construe '$_nFn$' "as a designation of the number correlated [via arithmetization] to the sentence resulting from the substitution of 'n' at the argument-place of the nth predicate in the syntactical denumeration of all the numerical predicates," but even so, Helmer argued, formula (1) "[could] not possibly be legitimated." (Helmer, 1937).

The articles of Grelling and Helmer were criticized in their turn by Rosser and Kleene in reviews published in the *Journal of Symbolic Logic* (Rosser (1938) and Kleene (1937b)). Rosser noted that "[Grelling's] exposition of the closing steps of Gödel's proof" did not agree with his own understanding of it, and that Grelling's statement quoted above to the effect that Q' was metamathematically undecidable was "indubitably false, ... since in a preceding sentence he says that Q' is a metamathematical theorem." Here, however, we may quibble: Grelling does *not* say that Q' is a metamathematical *theorem*—he refers to it merely as "ein metamathematischer Satz," which in this context simply means a statement or proposition; rather, Grelling's statement is "indubitably false" precisely because Q' *is* a metamathematical theorem (in the sense that metamathematical arguments show Q to be *true* in its intended interpretation). As to Helmer, Kleene agreed that the source of Perelman's errors lay in his failure to distinguish between formulas and their Gödel numbers, but he also noted Helmer's failure "to distinguish as consistently between ... *metamathematical statements* and *formal mathematical sentences* as between those and ... *syntactical numbers*" of the latter. At the same time he affirmed the basic legitimacy of Perelman's equivalence (1), "Helmer notwithstanding."

The controversy sparked by Perelman's paper exposes not only the fragility of the earlier acceptance of Gödel's results, but also the misunderstandings of those results by their would-be defenders. Thus Grelling and Helmer, while criticizing Perelman's confusion between object- and metalanguage, were themselves not always careful about syntactic distinctions. Grelling saw a metamathematical contradiction where none existed, and Helmer wrongly traced Perelman's error to a formula that is in fact provable (though not so easily as Perelman thought); while Rosser, by mistranslating "Satz," failed to recognize the real source of Grelling's misunderstanding. Only Kleene emerged from the debate untarnished ("clean").

For a more detailed account of Perelman's claims and their refutation, the reader may turn to Chapter 3, Section 5 of Ladrière's book (1957). That source also includes a discussion of two other, still later objections to Gödel's work by Marcel Barzin (1940) and Jerzy Kuczyński (1938). Neither of their challenges seems to have received much notice at the time, so I shall devote little attention to them here, except to note that their criticisms, unlike Perelman's, were based on Gödel's detailed formal proof rather than his informal introductory arguments; like Perelman, however, both Barzin and Kuczyński thought Gödel had discovered an antinomy. (In essence, Barzin confused formal expressions with their Gödel numbers, while Kuczyński, to judge from Mostowski's (1938) review, overlooked the formal antecedent "Wid(κ)" in Gödel's second theorem.)

5. Lingering Doubts

In 1939, the second volume of Hilbert and Bernays' *Die Grundlagen der Mathematik* appeared, in which, for the first time, a complete proof of Gödel's second incompleteness theorem was given. Whether because of its meticulous treatment of syntactic details or because of Hilbert's implied imprimatur, the book at last seems to have stilled serious opposition to Gödel's work, at least within the community of logicians.[10]

Outside that community it is difficult to assess to what extent Gödel's results were known, much less accepted or understood. It seems likely that many working mathematicians either remained only vaguely aware of them or else regarded them as having little or no relevance to their own endeavors. Indeed, until the 1970s, with the work of Matijasevic and, later, of Paris and Harrington, number theorists could (and often did) continue to regard undecidable arithmetical statements as artificial contrivances of interest only to those concerned with foundations.

Among the few non-logicians (at least to my knowledge) who took cognizance of Gödel's work in writings of the period was Garrett Birkhoff. In the first edition of his *Lattice Theory* (1940) he observed (p. 128) that "the existence of 'undecidable' propositions ... seems to have been established by Skolem [n.b.] and Gödel." In a footnote, however, he qualified even that tentative acceptance, noting that "Such a conclusion depends of course on prescribing all admissible methods of proof... [and] hence... should be viewed with deep skepticism." In particular, he believed that "Carnap [had] stated plausible methods of proof excluded by Gödel." In the revised edition (1948), the corresponding footnote (p. 194) is weakened to read "The question remains whether there do not exist perfectly 'valid' methods of proof excluded by [Gödel's] particular logical system"; but the reference to Skolem is retained in the text proper, and an added sentence (p. 195) declares that the proof of the existence of undecidable propositions "is however non-constructive, and depends on admitting the existence of uncountably many 'propositions,' but only countably many 'proofs'." Clearly, Birkhoff had not actually read Gödel's paper. His statements echo those of Zermelo, but he does not cite Zermelo's report.

Of still greater interest are the reactions of two philosophers of stature: Ludwig Wittgenstein and Bertrand Russell. Wittgenstein's well-known comments on Gödel's theorem appear in Appendix I of his posthumously published *Remarks on the Foundations of Mathematics* (extracted in English translation on pp. 431–435 of (Benacerraf and Putnam 1964)). Dated in the preface of the volume to the year 1938, they were never intended to be published and perhaps should not have been—but that, of course, is irrelevant to the present inquiry. Several commentators have discussed Wittgenstein's remarks in detail (see, e.g., the articles by A.R. Anderson, Michael Dummett, and Paul Bernays, pp. 481–528 of (Benacerraf and Putnam 1964)), and nearly all have considered them an embarrassment to the work of a great philosopher. Certainly it is hard to take seriously such objections as "Why should not propositions... of physics... be written in Russell's symbolism?"; or "The contradiction that arises when someone says 'I am lying' ... is of interest only because it has tormented people"; or "The proposition '*P* is unprovable' has a different sense afterwards [than] before it was proved" (Benacerraf and Putnam, 1964, pp. 432, 434). Whether some more profound philosophical insights underlie such seemingly flippant remarks must be left for Wittgenstein scholars to debate; suffice it to say that in Gödel's opinion, Wittgenstein "advance[d] a completely trivial and uninteresting misinterpretation" of his results. (Gödel to Abraham Robinson, 2 July 1973.)

As to Russell, two passages are of particular interest here, one published and one not, dating respectively from 1959 and 1963. The former, from *My Philosophical Development* (p. 114), forms part of Russell's own commentary on Wittgenstein's work:

> In my introduction to the *Tractatus*, I suggested that, although in any given language there are things which that language *cannot express* [my emphasis], it is yet always possible to construct a language of higher order in which these things can be said. There will, in the new language, still be things which it *cannot say* [my emphasis again], but which can be said in the next language, and so on *ad infinitum*. This suggestion, which was then new, has now become an accepted commonplace of logic. It disposes of Wittgenstein's mysticism and, I think, also of the newer puzzles presented by Gödel.

It might be maintained that in this passage Russell is making the same point that Gödel himself stressed in his (1930/31), that by passing to successively higher types one can obtain a transfinite sequence of formal systems such that the undecidable propositions constructed within each system are *decidable* in all subsequent systems. Contrary to Russell, however, Gödel emphasized that each of the undecidable propositions so constructed is already *expressible* at the lowest level.

The second passage is taken from Russell's letter to Leon Henkin of 1 April 1963:[11]

> It is fifty years since I worked seriously at mathematical logic and almost the only work that I have read since that date is Gödel's. I realized, of course, that Gödel's work is of fundamental importance, but I was puzzled by it. It made me glad that I was no longer working at mathematical logic. If a given set of axioms leads to a contradiction, it is clear that at least one of the axioms must be false. Does this apply to school-boys' arithmetic, and, if so, can we believe anything that we were taught in youth? Are we to think that $2 + 2$ is not 4, but 4.001? Obviously, this is not what is intended.
>
> . . .
>
> You note that we were indifferent to attempts to prove that our axioms could not lead to contradictions. In this Gödel showed that we had been mistaken. But I thought that it must be impossible to prove that any given set of axioms does *not* lead to a contradiction, and, for that reason, I had payed little attention to Hilbert's work. Moreover, with the exception of the axiom of reducibility which I always regarded as a makeshift, our other axioms all seemed to me luminously self-evident. I did not see how anybody could deny, for instance, that q implies p or q, or that p or q implies q or p.
>
> ... In the later portions of the book ... are large parts consisting of ... ordinary mathematics. This applies especially to relation-arithmetic. If there is any mistake in this, apart from trivial errors, it must also be a mistake in conventional ordinal arithmetic, which seems hardly credible.
>
> . . .
>
> If you can spare the time, I should like to know, roughly, how, in your opinion, ordinary mathematics—or, indeed, any deductive system—is affected by Gödel's work.

A curious ambiguity infects this letter. Is Russell recalling his bewilderment at the time he first became acquainted with Gödel's theorems, or is he expressing his continuing puzzlement? Is he saying that, intuitively, he had recognized the futility of Hilbert's scheme for proving the consistency of arithmetic but had failed to consider the possibility of rigorously *proving* that futility? Or is he revealing a belief that Gödel had in fact shown arithmetic to be *inconsistent*? Henkin, at least, assumed the latter; in response to Russell's closing request, he attempted to explain the import of Gödel's second theorem, stressing the distinction between incompleteness and inconsistency. Eventually a copy of Russell's letter made its way to Gödel, who remarked that "Russell evidently

misinterprets my result; however he does so in a very interesting manner . . . ."
(Gödel to Abraham Robinson, 2 July 1973.)

6. Conclusions

Insofar as it refers to the acceptance of Gödel's results by *formalists*, the received view appears to be correct. Gödel's proofs dashed formalist hopes, but at the same time they were most persuasive to those committed to formalist ideals. In other quarters, the incompleteness theorems were by no means so readily accepted; objections were raised on both technical and philosophical grounds. Especially prevalent were the views that Gödel's results were antinomial or were of limited generality. Cardinality restrictions, in particular, were often perceived to be responsible for the phenomenon of undecidability.

Gödel succeeded where others failed because of his attention to syntactic and semantic distinctions, his restriction to particular formal systems, and his concern for relative rather than absolute undecidability. He anticipated resistance to his conclusions and took pains to minimize objections by his style of exposition and by his avoidance of the notion of objective mathematical truth (which was nevertheless central to his own mathematical philosophy). Though aware of criticisms of his work, he shunned public controversy and considered his results to have been readily accepted *by those whose opinion mattered to him*; nevertheless, his later extensions of his results display his concern for establishing their generality.

In the long run, the incompleteness theorems have led neither to the rejection of formal systems nor to despair over their limitations, but, as Post foresaw, to a reaffirmation of the creative power of human reason. With Gödel's theorems as centerpiece, the debate over mind versus mechanism continues unabated.

Notes

1. Remarks, 14 March 1951, at the Einstein Award ceremonies.
2. "Schwierigkeit," not "Unmöglichkeit": We know from other sources that Gödel did not obtain his second theorem until after the Königsberg meeting.
3. Even earlier, on 22 October, Gödel (1930) had communicated an abstract of his results to the Vienna Academy of Sciences.
4. In a letter to Constance Reid of 22 March 1966, Gödel stated that he "never met Hilbert . . . nor [had] any correspondence with him." The stratification of the German academic system may have discouraged contact between the two men.
5. In the draft of a reply to a graduate student's query in 1970, Gödel indicated that it was precisely his recognition of the contrast between the formal definability of demonstrability and the formal *un*definability of truth that led to his discovery of incompleteness. That he did not bring this out in (1931) is perhaps explained by his observation (in a crossed-out passage from that same draft) that "in consequence of

the philosophical prejudices of [those] times ... a concept of objective mathematical truth ... was received with greatest suspicion and widely rejected as meaningless." For a fuller discussion of Gödel's avoidance of semantic issues, see Feferman (1985). Re Tarski's theorem on the undefinability of truth, see Tarski (1956), especially the bibliographical note, p. 152; footnote 1, pp. 247–248; the historical note, pp. 277–278; and footnote 2, p. 279. In those notes Tarski makes clear his indebtedness to Gödel's methods, relinquishing, so it would seem, any claim to priority for Gödel's own results (except for the prior exhibition of a consistent yet ω-inconsistent formal system).

6. Church, for example, in a letter to Gödel dated 27 July 1932, stated "I have been unable to see ... that your conclusions apply to my system" (as, in fact, they did not, since that system was subsequently shown to be inconsistent). In recent correspondence with the author, Church has acknowledged that he "was among those who thought that the ... incompleteness theorems might be found to depend on the peculiarities of type theory," a theory whose "unfortunate restrictiveness" he was trying to escape through a "radically different formulation of logic." Church goes on to remark, "indeed [the calculus of λ-conversion] might be claimed as a system of logic to which the Gödel incompleteness theorem does not apply. To ask in what sense this claim is sound and in what sense not is not altogether pointless, as it may give insight into ... where the boundary lies for applicability of the incompleteness theorem." (A. Church to J. Dawson, 25 July 1983.)

7. On his own offprint of Zermelo's remarks, Gödel marked this statement with the annotation "Dieser Behauptung ist mir ganz nicht verstandlich und scheint ... in sich widerspruchsvoll zu sein ... ."

8. Quotations and paraphrases of Post in this and the following paragraph are from his letter to Gödel of 30 October 1938.

9. It would be most interesting to know Gödel's opinion of Post's work. I have found no letters to Post in Gödel's *Nachlass* and little mention of Post in Gödel's other correspondence.

10. Not that Hilbert capitulated in the face of Gödel's results; for in the introduction to the first volume of *Grundlagen der Mathematik* (dated March 1934) Hilbert declared:

> Im Hinblick auf dieses Ziel [showing the consistency of arithmetic] möchte ich hervorheben, dass die zeitweilig aufgekommene Meinung, aus gewissen neueren Ergebnissen von Gödel folge die Undurchführbarkeit meiner Beweistheorie, als irrtümlich erweisen ist. Jenes Ergebnis zeigt in der Tat auch nur, dass man für die weitergehenden Widerspruchsfreiheitsbeweise den finiten Standpunkt in einer schärferen Weise ausnutzen muss, als dieses bei der Betrachtung der elementaren Formalismen erforderlich ist.

11. I am grateful to the Russell Archives, McMaster University, for permission to quote from this letter.

References

Barzin, M. (1940), Sur la Portée du Théorème de M. Gödel, *Académie Royale de Belgique, Bulletin de la Classe des Sciences*, Series 5, **26**, 230–239.

Benacerraf, P. and Putnam, H. (eds.) (1964), *Philosophy of Mathematics: Selected Readings*, Prentice-Hall, Englewood Cliffs, NJ.

Church, A. (1946), Review of Finsler (1944), *J. Symbolic Logic* **11**, 131–132.

Davis, M. (1965), *The Undecidable*. Raven Press, Hewlett, NY.

Davis, M. (1982), Why Gödel didn't have Church's thesis, *Inform. and Control* **54**, 3–24.

Dawson, J.W., Jr. (1984), Discussion on the foundation of mathematics, *Hist. Philos. Logic* **5**, 111–129.

Dawson, J.W., Jr. (1985), Completing the Gödel–Zermelo correspondence, *Historia Math.* **12**, 66–70.

Feferman, S. (1985), Conviction and caution: A scientific portrait of Kurt Gödel, *Philos. Natur.* **21**, 546–562.

Finsler, P. (1926), Formale Beweise und die Entscheidbarkeit, *Math. Z.* **25**, 676–682. (English translation in van Heijenoort (1967), pp. 440–445.)

Finsler, P. (1944), Gibt es unentscheidbare Sätze? *Comment. Math. Helv.* **16**, 310–320.

Gödel, K. (1930), Einige metamathematische Resultate über Entscheidungsdefinitheit und Widerspruchsfreiheit, *Anzeiger der Akademie der Wissenschaft in Wien* **67**, pp. 214–215. (English translation in van Heijenoort (1967), pp. 595–596.)

Gödel, K. (1930/31), Über Vollständigkeit und Widerspruchsfreiheit, *Ergebnisse eines mathematischen Kolloquiums* **3**, pp. 12–13. (English translation in van Heijenoort (1967), pp. 616–617.)

Gödel, K. (1931), Über formal unentscheidbare Sätze der Principia Mathematica und verwandter Systeme I, *Monatsh. Math. Phys.* **38**, 173–198. (English translation in van Heijenoort (1967), pp. 596–616.)

Gödel, K. (1934), On Undecidable Propositions of Formal Mathematical Systems, Mimeographed notes by S.C. Kleene and J.B. Rosser of lectures by Kurt Gödel at the Institute for Advanced Study. (As reprinted in Davis (1965), pp. 39–74.)

Gonseth, F. (ed.) (1941), *Les Entretiens de Zürich sur les Fondements et la methode des sciences mathématiques, 6–9 Decembre 1938*, Leeman, Zurich.

Grattan-Guinness, I. (1979), In memoriam Kurt Gödel: His 1931 correspondence with Zermelo on his incompletability theorem, *Historia Math.* **6**, 294–304.

Grattan-Guinness, I. (1981), On the development of logics between the Two World Wars, *Amer. Math. Monthly* **88**, 495–509.

Grelling, K. (1937/38). Gibt es eine Gödelsche Antinomie? *Theoria* **3**, 297–306. *Zusätze und Berichtigungen*, ibid. **4**, 68–69.

Hahn, H. et al. (1931), Diskussion zur Grundlegung der Mathematik, *Erkenntnis* **2**, 135–151. (English translation in Dawson (1984), pp. 116–128.)

Helmer, O. (1937), Perelman *versus* Gödel, *Mind* **46**, 58–60.

Kleene, S.C. (1937a), Review of Perelman (1936), *J. Symbolic Logic* **2**, 40–41.

Kleene, S.C. (1937b), Review of Helmer (1937), *J. Symbolic Logic* **2**, 48–49.

Kleene, S.C. (1976/78), The Work of Kurt Gödel, *J. Symbolic Logic* **41**, 761–778. Addendum, ibid. **43**, 613.

Kleene, S.C. (1981a), Origins of recursive function theory, *Ann. Hist. Comput.* **3**, 52–67. (Corrigenda in Davis (1982), footnotes 10 and 12.)

Kleene, S.C. (1981b), The theory of recursive functions, approaching its centennial, *Bull. Amer. Math. Soc.* (*N.S.*) **5**, 1, 43–61.

Kreisel, G. (1979), Review of Kleene (1978). (Addendum to Kleene (1976).) *Zentralblatt für Mathematik und ihre Grenzgebiete* **366**, 03001.

Kuczyński, J. (1938), O Twierdzeniu Gödla, *Kwartalnik Filozoficzny* **14**, 74–80.

Ladrière, J. (1957), *Les Limitations Internes des Formalismes*, Nauwelaerte, Louvain; Gauthier-Villars, Paris.

Menger, K. (1978), *Selected Papers in Logic and Foundations, Didactics, Economics*. (*Vienna Circle Collection*, Number 10.) Reidel, Dordrecht.

Mostowski, A. (1938), Review of Kuczyński (1938), *J. Symbolic Logic* **3**, 118.

Perelman, C. (1936), L'Antinomie de M. Gödel, *Académie Royale de Belgique, Bulletin de la Classe des Sciences*, Series 5, **22**, 730–736.

Popper, K.R. (1980), Der wichtigste Beitrag seit Aristoteles, *Wissenschaft aktuell* **4/80**, 50–51.

Post, E.L. (1965), Absolutely unsolvable problems and relatively undecidable propositions: Account of an anticipation, in Davis (1965), pp. 338–433.

Quine, W.V. (1978), Kurt Gödel (1906–1978). *Year Book of the American Philosophical Society* **1978**, 81–84.

Rosser, J.B. (1936), Extensions of some theorems of Gödel and Church, *J. Symbolic Logic* **1**, 87–91. (Reprinted in Davis (1965), pp. 231–235.)

Rosser, J.B. (1938), Review of Grelling (1937/38), *J. Symbolic Logic* **3**, 86.

Tarski, A. (1956), *Logic, Semantics, Metamathematics*. Edited and translated by J.H. Woodger. Oxford University Press, Oxford.

Turing, A.M. (1936/37), On computable numbers, with an application to the Entscheidungsproblem, *Proc. London Math. Soc.*, Series 2, **42**, 230–265. Corrigenda, ibid. **43**, 544–546. (Reprinted in Davis (1965), pp. 115–154.)

van Heijenoort, J. (ed.) (1967), *From Frege to Gödel. A Source Book in Mathematical Logic, 1879–1931*, Harvard University Press, Cambridge, MA.

Wang, H. (1981), Some facts about Kurt Gödel, *J. Symbolic Logic* **46**, 653–659.

Zermelo, E. (1932), Über Stufen der Quantifikation und die Logik des Unendlichen, *Jahresber. Deutsch. Math. Verein.* **41**, part 2, 85–88.

Gödel's and Some Other Examples of Problem Transmutation

Hao Wang

I am much struck by the historical fact that in attempting to prove the (relative) consistency of analysis, Gödel was led to certain propositions undecidable in segments of set theory. He then brought the propositions into an elegant arithmetic form and, independently of this improvement, realized that his result also implied a negative reply to the initial problem of proving consistency (not only of analysis but even of number theory). For lack of a suitable term, I have chosen to call this phenomenon "problem transmutation," by which I mean to cover an indefinite and wide range of related phenomena.

I am aware that I do not have a precise concept here, but wish merely to illustrate the unexpected course of one task leading to another, by assembling a few examples with which I happen to be familiar. Besides more examples from Gödel, I shall include my less controlled movement from automatic theorem proving to the invention of a combinatorial game which was meant to help give a decision procedure for, but instead made possible a negative solution of, an interesting case of the Entscheidungsproblem. In addition, a conjecture is offered in regard to the influence of this type of work on S.A. Cook's surprising discovery on the complexity question of the propositional calculus.

1. Historical Background: Hilbert and Proof Theory

In 1900 Hilbert turned his attention from the foundations of geometry to offer an axiom system for analysis, departing from the usual genetic approach to the study of the concept of (real) number (in "über den Zahlenbegriff"). Indeed the consistency of some such system is listed as the second problem in the list of his Paris lecture of the same year. Hilbert's axioms characterize the real numbers as forming a real Archimedean field which permits no further extension to any field of the same kind.

The requirement of nonextendibility is given in a pregnant axiom of completeness which, however, has a complex logical structure and is indeed somewhat ambiguous. As we now know, this matter of closing-off or completeness calls for going beyond formal systems even when we consider the simpler case of integers. In current terminology, only the second-order Peano axioms are complete, first-order axioms for the integers are incompletable. It is interesting to see that Hilbert soon came to relate the question of axiomatically closing-off the theory of integers to that of the indeterminate range of what makes up of logic.

In 1901 he gave a lecture to the Göttingen mathematical society on the problems of completeness and decision (of true or false by the axioms). In E. Husserl's formulation, Hilbert asked (*Philosophie der Arithmetik*, Husserlina, 1970, p. 445): "Would I have the right to say that every proposition dealing only with the positive integers must be either true or false on the basis of the axioms for positive integers?" "When we assert that a proposition is decided on the basis of the axioms of a domain, what may we use for this apart from the axioms? The whole of logic. What is that? All theorems, which are free from every particularity of a domain of knowledge, what holds independently of all 'special axioms,' of all material of knowledge.—But we come thus to a neat fix: in the domain of algorithmic logic? in that of finite cardinals? in the theory of combinations? in the general theory of ordinals numbers? And finally isn't the richest set theory itself purely logical?"

I have quoted this passage in full because it is relevant to certain current debates, particularly with regard to the range of logic. I tend to agree with Dedekind and Hilbert (and Gödel) that logic includes set theory. If we use this concept of logic, then we are justified in accepting the familiar argument for proving Peano's axioms categorical. Hence, each proposition is "decided" by the axioms (i.e., either true or false). But this decidability or determination is different from the more strict requirement of an algorithm to carry out the decision in each case. The stricter decidability is the concept which is familiar nowadays and most likely also what Hilbert wanted. But Hilbert's foremost concern was consistency proofs; next came completeness which for him was probably a weaker requirement in general than decidability.

In his 1904 Heidelberg lecture on the foundations of logic and arithmetic Hilbert for the first time observed that, while one can prove the consistency of geometry by an arithmetic interpretation, for the consistency of arithmetic (meaning number theory and analysis) "the appeal to another fundamental discipline seems unreasonable." He suggested a plan of building up logic and arithmetic simultaneously, as well as the thought of translating proofs into the language of formulas (of symbolic logic) thereby turning the proof of consistency (the proof of the "existence of the infinite") into a problem of elementary arithmetic. Probably for lack of good ideas and for interest in other matters (particularly integral equations and physics), Hilbert did not return to this area until his 1917 Zurich lecture on axiomatic thinking in which Hilbert praised the "axiomatization of logic" by Whitehead and Russell as "the crowning of all the work of axiomatization." Of course there remained for Hilbert the matter of proving consistency and he proposed that the concept of specific mathematical proof be itself made the object of study.

It was only since 1920 that Hilbert, in collaboration with Paul Bernays and provoked by the opposition from Weyl and Brouwer, began to concentrate on proof theory. The fruits of this labor were reported in the first Hamburg lecture of 1922, the 1922 Leipzig lecture, the 1925 Münster lecture on the continuum, and the second Hamburg lecture of 1927. Meanwhile, apart from Bernays, other young mathematicians, notably M. Schönfinkel, W. Ackermann, and J. von Neumann, also joined in the exciting enterprise. By 1928 it was believed that the consistency of number theory had been achieved by the finitist method envisaged by Hilbert (and Bernays) and that even a similar proof of the consistency of analysis had been essentially completed, only "the proof of a purely arithmetical elementary theorem of finiteness" was wanting.

2. Hilbert's List of Problems

On 3 September 1928 Hilbert delivered the address "Probleme der Grundlegung der Mathematik" and he "received a stormy applause at the beginning as well as after the end of the address." [*Atti del Congresso Internationale dei Mathematici, Bologna* 3–10 *Septembre* 1928, Bologna, 1929, vol. I, pp. 135–141; also *Math. Ann.*, vol. 102 (1930), pp. 1–9; also *Grundlagen der Geometrie*, seventh edition, 1930, Anhang X; also K. Reidemeister, *Hilbert Gedenkband*, 1971, pp. 9–19].

Hilbert began his address by reviewing general advances in mathematics over the 10-year period to 1928 and then announced that the work of Ackermann and von Neumann had carried out a (finitist) consistency proof of number theory. Then he listed four yet unresolved problems.

PROBLEM 1 (Finitist) consistency proof of the basic part of analysis (or second-order functional calculus). Hilbert stated that Ackermann had already worked out the main part of such a proof, only an elementary finiteness condition remained to be established. He remarked that the proof would justify the theory of real numbers with the Dedekind cuts, the second-order Peano axioms, and the theory of Cantor's second number class.

PROBLEM 2. Extension of the proof to higher order functional calculi (or the part of the simple theory of types to three or four or five levels). Hilbert mentioned in this connection the additional problem of proving the consistency of a stronger axiom of choice, because his formulation contains a certain form of the axiom of choice as an integral part of the basic system. It is interesting that Hilbert did not mention explicitly full set theory but confined himself to what we would now consider small fragments of set theory adequate to the development of standard mathematics (of 1928).

PROBLEM 3. Completeness of the axiom systems for number theory and analysis. Hilbert remarked on the familiar fact that second-order theories are indeed complete in both cases on the basis of the usual proofs that they are categorical. The problem is to render the proofs of categoricalness "finitely rigorous." For number theory, Hilbert explained his idea by requiring that the proof of isomorphism of two models be "finitely recast" to show that whenever a proposition A can be shown to be consistent with the axioms the opposite of A cannot be shown to be so. In other words, whenever A is consistent with the axioms, it is provable. This seems to be more specific than the familiar idea of so rendering the second-order axioms more explicit that we do have "formal" axiomatic systems. In any case, Hilbert appears to believe at that time that such complete formal systems exist.

Hilbert did not mention the question of decidability of number theory and analysis. We are now familiar with Turing's observation that if a formal system is complete in the sense just mentioned it is also decidable. If Hilbert had been aware of this connection at that time, he would probably have been more skeptical about the existence of complete formal systems in the two cases, since presumably he would have more doubt about these theories being decidable.—But I am merely conjecturing and have not looked into the historical evidence about Hilbert's beliefs at that time on these questions of completeness and decidability.

PROBLEM 4. The problem of completeness of the system of logical rules (i.e., the first-order logic) in the sense that all (universally) valid sentences are provable. Hilbert offered this as an example of purely logical problems which fall under his proof theory. This same problem was also mentioned in the first edition of the Hilbert–Ackermann textbook of 1928.

Hilbert did not list the Entscheidungsproblem (for the first-order logic), which was of course a central problem in his school, except to mention that for monadic quantification theory the known method of solving the Entscheidungsproblem (Schröder's elimination problem) also yielded a proof of completeness of the rules in the subdomain. Perhaps Hilbert did not include this or the decision problem for number theory because he did not regard them as belonging to his proof theory.

3. Gödel's Dramatic Response

In the summer of 1929, after graduating from the University of Vienna, Gödel began to concentrate on research. At about this time he read the Hilbert–Ackermann textbook. Gödel soon settled Hilbert's Problem 4 by proving the completeness of the first-order logic, and the whole work had been written up as his dissertation and approved by autumn 1929.

In the summer of 1930 Gödel undertook to study the problem of proving the consistency of analysis. He was probably familiar with the work of the Hilbert school and aware that there were gaps in the proposed finitist consistency proof of number theory. In his own words of 1976 as he had me formulate in the third person:

> He found it mysterious that Hilbert wanted to prove directly the consistency of analysis by finitist method. He believes generally that one should divide the difficulties so that each part can be overcome more easily. In this particular case, his idea was to prove the consistency of number theory by finitist number theory, and prove the consistency of analysis by number theory, where one can assume the truth of number theory, not only the consistency. The problem he set for himself was the relative consistency of analysis to number theory; this problem is independent of the somewhat indefinite concept of finitist number theory.

I am under the impression that according to the consensus today, finitist number theory as originally envisaged by Hilbert consists basically of the quantifier-free theory of primitive recursive functions. It is on the basis of an identification more or less like this that we can speak of Gödel's results as settling Hilbert's problems unambiguously. Gödel's contrast between consistency and truth of number theory calls for some comment. As I understand his distinction, it allowed for different answers to the then unsettled question whether there exists complete formal systems of number theory. If there were such a system, then its consistency would also give truth of number theory. Otherwise the remaining task of proving the consistency of number theory by finitist number theory would seem to call for some clarification. One possibility is to envisage a whole class of formal systems of number theory which taken together would determine truth in number theory in the strong sense that all and only true propositions of number theory are provable in the union of the class, as well as a scheme of proving the consistency of each of the systems by finitist number theory. Another possibility is to prove by finitist number theory the consistency of a standard though incomplete system of number theory, say the first-order number theory, which, though incomplete, determines univocally truth in number theory, e.g., by containing an adequate definition of truth in number theory. Given the distinction of consistency from truth, there is an idle question of adding something to help bring together the two halves envisaged by Gödel, in order to yield the stronger conclusion of a consistency proof of analysis by finitist method. Gödel undoubtedly found truth in number theory much clearer than the consistency of analysis so that the problem he set for himself was both important and more definite than the other half.

Gödel quickly found that truth in number theory cannot be defined in number theory, quite apart from particular formal axiom systems for number theory. He was then able to formulate propositions in elementary segments of type theory or set theory which are not decidable in them. After announcing the result in the September meeting at Königsberg he improved the result to obtain elegant arithmetic propositions which are undecidable in the given formal system, and also discovered that for the familiar systems of number theory and set theory, there can be no consistency proof of any of these systems formalizable in the system itself. Hence, since the finitist method as explained by Hilbert should be formalizable in these systems, Gödel had not only settled Hilbert's Problems 1 and 2 negatively, but also refuted Hilbert's belief that a finitist consistency proof of number theory had already been found. Moreover, he had also settled Hilbert's Problem 3 negatively. In all these cases Gödel proved that the finitist proofs and complete formal systems asked for in these questions simply cannot exist.

In the spring of 1961 I had an opportunity to go with a friend to visit L.E.J. Brouwer at his house (uninvited as was the custom with him). He did most of the speaking and discoursed on diverse topics apparently at random. I remember his mentioning, for totally different reasons in each case, Huygens, Hadamard, Heyting, and Gödel. Of Gödel's incompleteness results, he expressed astonishment that so much had been made of them, saying that the conclusions had been evident to him for a long time before 1931. Of course a major difference is that Gödel's precise formulations were convincing to people with widely different beliefs while Brouwer's convictions were based on a motley of much less explicitly or homogeneously expressed interesting ideas. Moreover, the wide implications of Gödel results can only be negated from a highly special point of view that to most mathematicians and philosophers must appear highly arbitrary and one-sided.

Even though it seems fair to say that Gödel had destroyed Hilbert's original program, Gödel continued to be interested in the underlying problem of mathematical evidence, and to look for natural broadenings of Hilbert's stringent requirements. In 1933 Gödel published a note to establish that classical number theory is as consistent as intuitionistic number theory (an observation also made independently by Bernays and by G. Gentzen). In 1942 Gödel found an interpretation of intuitionistic number theory by primitive recursive functionals which were regarded by him as an extension of Hilbert's finitist method.

Another "transmutation" of problems and ideas is in connection with the continuum problem. In 1976 Gödel thought that it was in 1930 when he began to think about the continuum problem and heard about Hilbert's proposed outline of a proof of the continuum hypothesis in axiomatic set theory as presented in the 1925 lecture "On the infinite." Hilbert's idea was to just construct (recursive) ordinals and use recursively defined sets and functions of orders indexed by such ordinals. Consistent with his belief that finitist consistency proofs exist for strong systems, Hilbert believed and thought he could prove that these exhausted all the ordinals and all the sets of integers. Hence, since it is easy to see and prove in set theory that the cardinalities of these ordinals and these sets are the same, he thought that he had (the outline of) a proof of the continuum hypothesis in set theory. More, Hilbert thought that the axioms of set theory are also true in his constructive interpretation along the same line. That is why he thought he also had (the outline of) a finitist consistency proof of the axioms of set theory (as known then).

Gödel used the idea of analyzing sets (of integers) into orders indexed by ordinals. Instead of using recursively defined sets, he soon thought of the first-order definable sets in the ramified hierarchy. He experimented with different ways of introducing (countable) ordinals for some time. Finally the idea of using all ordinals occurred to him. This implies that instead of proving the continuum hypothesis one could only hope for a relative consistency proof unless one could get all ordinals in some suitably explicit way. In 1976 Gödel thought that it was in 1935 when he verified that all the axioms of set theory at the time, including the axiom of choice, are satisifed by the constructible sets, i.e., the ramified theory over all ordinals. In 1938 he proved that the (generalized) continuum hypothesis also holds in this model.

4. The Entscheidungsproblem

The practice of speaking of the decision problem of the first-order logic as *the* Entscheidungsproblem must have begun around 1920. For example, this usage is in the title of H. Behmann's paper of 1922 (in *Mathematische Annalen*). Gödel's report in 1932 of his decision procedure for an interesting special case also follows this usage. In the literature the first proof of the unsolvability of the general case is commonly attributed to A. Church. But it could be argued, and I believe this is Gödel's view, that A.M. Turing was really the first one to give a complete proof of this result. This is because it was Turing who for the first time gave a convincing explication of the intuitive concept of computability and algorithm. Hence, only after this explication can we hope to get general unresolvability results. From this viewpoint Turing's own proof of the unsolvability of the Entscheidungsproblem is directly complete while Church's slightly earlier proof is only completed by adding Turing's explication plus the proof that all Turing computable functions are indeed recursive.

From 1953 on I began to think about the relation between logic and computers. For a long time before that I had felt disappointed that the formal precision in making proofs entirely explicit seems pointless in practice and most mathematicians have little sympathy with this aspect of logic. I was, therefore, pleased with the idea that fast computers can take advantage of the precision to take over parts of the task of proving theorems, filling in gaps, etc. Some of these general thoughts were written up in 1954 and included as additional remarks in a paper devoted to a programming version of Turing machines, later published at the beginning of 1957 in *Journal of the Association for Computing Machinery*. It was not until the summer of 1958 that Brad Dunham provided me with an opportunity to learn programming and carry out in part these vague ideas. The result was surprisingly gratifying not only in the quick speed but especially in the unexpected discovery that all the theorems of *Principia Mathematica* in first-order logic fall under a very simple and familiar decidable subclass with fast decisions. These results were written up and submitted by December 1958 but for some curious reason the IBM *Journal* published it only in January 1960.

Meanwhile I had obtained in July 1959 a visiting position at Bell Laboratories to continue this work. After tidying up the loose ends left over from summer 1958, I began to investigate the patterns of Herbrand expansions of natural classes of formulas in first-order logic. I came upon the major unsettled case of *AEA* formulas, i.e., formulas of the simple form $AxEyAzM$-xyz, with no quantifiers in M. Most people who had worked on it since the 1920s probably conjectured that the case is decidable. The patterns of Herbrand expansions of these formulas are especially simple and can be represented in a natural way by certain simple finite sets of points on the Cartesian plane. Since my colleagues at Bell Labs were not familiar with logic, I was frustrated in trying to discuss with them what I thought was an elegant combinatorial problem. In the beginning of 1960 I came upon a picturesque formulation of a closely related problem, known since as the domino problem or Wang tiles.

At that time I thought that the basic domino problem must be decidable (i.e., every solvable set has a periodic solution) and that the decision procedure would lead to one for the *AEA* case. Soon I realized that if some domino or tile is required to occur in every solution (the "origin-constrained" problem), then one can simulate Turing machines and thereby show the modified problem to be unsolvable. More than a year later I lectured at Harvard on these ideas (in autumn 1961) and within a month, with the collaboration of a student, A.S. Kahr, who had no previous training in logic at all, I was able to show that my construction can be extended from the origin-constrained problem to the "diagonal-constrained" problem, thereby showing that the *AEA* case is, contrary to expectation, unsolvable and indeed a reduction class for the whole of first-order logic. This incidentally unifies all the reduction classes in a simple way. About two years later another student, Robert Berger, showed that the basic (i.e., unconstrained) domino problem is also unsolvable, contrary to my original conjecture. These dominoes or colored tiles have found various other applications over the last 20 years or so.

The proof of the unsolvability of the *AEA* case by colored tiles also goes through the simulation of Turing machines and can be viewed as a drastic simplification of Turing's proof in the sense that he used much more complex formulas than those in the *AEA* form to express basically the same thing about Turing machines. Steve Cook came around a little later and was before long fully indoctrinated in these economical representations. With his intense interest in computational complexity, he was a few years later able to carry the economy even further by using the idea of indeterministic computation which is to deterministic computation much as provability is to decidability. He proved in his paper of 1971 (ACM Conference on Theory of Computing) that nondeterministic Turing machines can even be represented by formulas in the propositional calculus in such a way as to sustain a desired relation between the consistency (or satisfiability) of a formula and certain specific behaviors of its corresponding machine. This unexpected "transmutation" has led to the formulation of the central open problem of computational complexity, commonly stated in the abbreviated form of asking whether $P = NP$.

The Development of Self-Reference: Löb's Theorem

C. Smoryński

Social processes operate on the sciences and social history must be understood in order to understand the history of science. This simple truth has transcended mere axiomatic truth; it has transcended mere fashion; today, the social treatment is *de rigueur* in the history of science and the history of computing. In the history of mathematics, however, the social emphasis is not so firmly entrenched. Historians who discuss the history of mathematics certainly steep their discussions in social contexts, but the history of mathematics is quite often written by mathematicians or, at least, mathematical educators, and mathematicians generally do not care about the overall societal background. Modern mathematics has largely divorced itself from the sciences and has become something of an intellectual *sport*, a game of ideas. Now, mathematicians are somewhat social and love anecdotes about fellow mathematicians, but this is as far as their interest in social history goes: Where the sciences inject history into their textbooks and discuss the growth of ideas, the history in most calculus texts limits itself to paragraph-length biographies, and one of the most successful textbooks on the history of mathematics is highly anecdotal, its author having additionally published no fewer than three books consisting of nothing but anecdotes. When the mathematician wants to discuss history (i.e., history of mathematics) and he wants to discuss it seriously and will not settle for mere anecdotes, it is not the social forces that created a need for a given type of mathematics that he writes about—mathematics has the odd habit of developing long *before* it is useful (i.e., before it is needed by society at large); it is not a simple chronology that he lists—despite all one reads about how the mathematician is like a fish out of water when he is not doing mathematics, he is not as stupid as all that; it is the development of the subject that interests him: What was the key idea? How did it come about? etc.

The topic of the present paper is the development of self-reference in, say, arithmetic, i.e., formal linguistic self-reference of the kind used by Gödel as opposed to the functorial self-reference of the Recursion Theorem. The most salient feature of this opposition is the paucity of the former as compared with the opulence of the latter. How can we account for this? An explanation can occur on one of many levels. One could surmise that formal linguistic self-reference interests the philosophically minded logician, that the mathematically oriented logician would be attracted to the functorial self-reference, and that, as mathematicians are more inclined toward technical development, we would naturally expect greater development of the Recursion Theorem and related recursion theoretic matters. The socially minded historian could even point to references to computability in the titles of books on recursion theory and, inferring some relationship, emphasize the increasing importance of computers to society. Put so bluntly, he would be wrong, but a knowledgable scientific historian—with a knowledge of philosophy—could make a good case for something similar: Recursion Theory—all the undecidability results, complexity results, and results generally revealing our mathematical limitations—is the natural mathematical expression of the central rôle of intellectual impotence of twentieth century Western thought. (I refer here to du Bois-Reymond's *ignorabimus*, relativity theory, the Heisenberg uncertainty principle, logical positivism, and—eventually—Gödel's Theorems and their wake; if I knew anything about it, I am sure I could list trends in art and literature as well.) Such a theory, like Spengler's *Decline of the West*, is very pleasing and gives one a very broad perspective, but interpreting it—if possible at all—is more of an art—like divination—than a technology, and this is, to me, a failing of all social history of the sciences: It is painting with broad strokes; it gives a good background, but doesn't do too well with details.

I wish in this chapter to be very specific. I also wish to be mathematical, and I mean this in two ways: First, and rather obviously, I intend to be mathematical in historical style; I shall look at the *mathematical* development. Second, I take seriously the existence of subjective phenomena, something not done by all social historians, whose belief in objectivity I cannot swallow. (An extreme example is given by Derek de Solla Price. Incidentally, I am here overexaggerating the anti-subjectivity of the social historian: Goldberg's *Understanding Relativity* is a good refutation of my exaggeration. Nonetheless, there is something to my remarks.)

The question I wish to address (not directly, rather it is to be kept in the back of one's mind) is: How do we account for the slow development of formal linguistic self-reference, particularly when we compare it to the development of recursion theory? In 'Fifty years of self-reference in arithmetic' (Smoryński, 1981b), I suggested that most mathematical logicians found Kleene's recursion theory an emotionally safer outlet for their self-referential propensities. This is another exaggeration, but, again, one with some foundation: How many textbooks do not draw a distinction between mere incompleteness (*some* sentence is undecided) and Gödel's First Incompleteness Theorem (the sentence asserting its own unprovability is undecided)? Are not modern mathematical logicians openly embarrassed at the blatant intensionality behind Gödel's two Incompleteness Theorems? (Cf., e.g., the opening paragraphs of Friedman, McAloon, and Simpson (1982) and Simpson (1983) in this regard; note, incidentally, how the authors of the former decry the artificiality of the consistency statement proven undecided by Gödel while simultaneously pointing with pride in the title of their paper to the equivalence of their undecidable sentence with 1-consistency!) Also, in Smoryński (1981b), I noted that so basic a result as the Diagonalization Lemma, which is a clear antecedent to Kleene's fully generally formulated Recursion Theorem of 1938, was not stated in any form in many textbooks. This, I said, was another example of the negative attitude logicians had (and have) toward formal linguistic self-reference.

There is something to the above explanation, but it is, nevertheless, not completely satisfying. Not only is it hopelessly naïve, but, like the application of social history to anything specific, it is too broad, too general, and cannot be accepted as an explanation. In discussing microcosmic developments, one prefers microcosmic to macrocosmic explanations. Moreover, the mathematician wants a mathematical explanation, preferably a theorem to the effect that A had to precede B (say), but an explanation in terms of mathematical practice will do, e.g., A could not have done B because everyone was thinking C and the goal D would not have required B .... In these terms, Kuhn's language of paradigms and paradigm shifts seems just the thing the mathematician might like. Why did Recursion Theory develop instead of Proof Theory? *Answer*: A paradigm shift! Kleene's success was so great it became the new paradigm; incompleteness gave way to undecidability. I shall use this language and try to keep in focus what researchers had in view in the sequel. My topic will be the very microcosmic one of the history of Löb's Theorem.

1. Henkin's Problem and Kreisel's First Response

In the 1950s, the tie between logic and philosophy was stronger than it is today. Where nowadays the textbooks are invariably titled *Introduction to Mathematical Logic* and emphasize model theory and recursion theory, the textbook of the 1950s was Kleene's *Introduction to Metamathematics* and it had an extended discussion of foundations and the paradoxes. The paradoxes, needless to say, were the first paradigm in self-reference. I have the impression that interest in the paradoxes survived well into the 1950s. Incompleteness results were still partially viewed as analogues to paradoxes and variants of the paradoxes were transformed into variant incompleteness results. Gödel's self-referential sentence,

"I am unprovable,"

was a formal counterpart to the paradoxical

"I am lying."

As with the latter, if "I am unprovable" is false, it must be provable, whence true—whence unprovable; but nothing paradoxical now occurs. The sentence

"I am telling the truth,"

doesn't lead to any paradox on assumption of its truth or falsity. There seems to be no way of deciding whether it should be true or false. By analogy, one could not readily decide whether

"I am provable"

should be true or false. Since the truth of the Gödel sentence is decidable, it seemed plausible to Leon Henkin that someone could decide this new sentence.

In 1952, the *Journal of Symbolic Logic* instituted a short-lived problems column. Herein Henkin posed the problem of the truth or falsity of the sentence "I am provable." His exact wording, which is important in understanding the sequel, was:

> If Σ is any standard formal system adequate for recursive number theory, a formula (having a certain integer q as its Gödel number) can be constructed which expresses the proposition that the formula with Gödel number q is provable in Σ. Is this formula provable or independent in Σ?

Nowadays, we would state this problem as follows:

Let $Pr(v)$ be the formula expressing provability in the system Σ referred to above. Gödel produced a sentence φ satisfying

$$\Sigma \vdash \varphi \leftrightarrow \neg Pr(\ulcorner\varphi\urcorner),$$

(where $\ulcorner\varphi\urcorner$ is the Gödel number of φ) and showed that φ is unprovable. Suppose instead that a sentence φ satisfies

$$\Sigma \vdash \varphi \leftrightarrow Pr(\ulcorner\varphi\urcorner).$$

Is φ provable?

This is *not* an accurate rendering of Henkin's problem, but is, rather, the generalization affirmatively solved by Löb. Kreisel's response fairly dramatically points to the difference between the two formulations.

In a note, Georg Kreisel (1953) showed that the answer to Henkin's problem was "yes" and "no"; it depended on how one expressed provability and on the formula constructed to express its own provability. The key word here, and in Henkin's formulation of the problem, is "express."

The Diagonalization Lemma in its most natural *mathematical* formulation, if not its strongest such, asserts that to any arithmetic formula ψv, with the one free variable v, there is a sentence φ such that, retaining Henkin's Σ,

$$\Sigma \vdash \varphi \leftrightarrow \psi(\ulcorner\varphi\urcorner).$$

In Smoryński (1981b), I pointed out that many logic textbooks do not state this lemma and, instead, repeat the construction every time the authors wish to apply the lemma. I exhibited this as an instance of the lack of a genuine development of self-reference. This is, I think, the case: Mathematically, it makes good sense to prove the lemma and apply the provable equivalence of φ to $\psi(\ulcorner\varphi\urcorner)$—provided one knows, as experience teaches us, that this is generally all one needs to know about φ and its relation to ψv. If one has not studied self-reference, one merely follows tradition .... In any event, this mathematical experience was not available in 1953 when Kreisel's paper was submitted, nor, *a fortiori*, in 1952 when Henkin posed his problem. Moreover, there is the philosophical issue: The relation between the formula ψv and the fixed point φ constructed by Gödel's method is stronger than the mere equivalence of φ and $\psi(\ulcorner \varphi \urcorner)$; φ literally has the form $\psi(t)$ for some term t which evaluates to the code $\ulcorner \varphi(t) \urcorner$. Thus, φ literally expresses $\psi(\ulcorner \varphi \urcorner)$. Henkin did not want to know if some sentence accidentally equivalent to the assertion of its own provability was provable (I assume the truth of this, that

$$\Sigma \vdash \bar{0} = \bar{0} \leftrightarrow Pr(\ulcorner \bar{0} = \bar{0} \urcorner),$$

was obvious); he did not ask if *all* such sentences were provable (there being no reason at the time to believe in—or even surmise—such uniformity); he wanted to know if the (a?) sentence so *constructed* to *express* its own provability was provable. It was this question Kreisel addressed.

Kreisel relaxed slightly the stricture that φ be *constructed* to express its own provability. To him, for φ to express that it had a property ψv, it was enough for φ to be of the form $\psi(t)$ for some term t evaluating to $\ulcorner \varphi \urcorner$; t did not have to be the term constructed by Gödel's method. There is no evidence that Henkin or anyone objected to this relaxation. Kreisel departed more radically from Henkin's intended problem by choosing his own means of expressing provability rather than using the predicate $Pr(v)$ constructed by Gödel. Should he have stuck to Gödel's predicate? What does, or should, one mean by saying that a predicate expresses provability? Kreisel proposed the following:

Criterion for the Expression of Provability. A formula $P(v)$, with only v free, expresses provability in the system Σ iff, for any sentence φ, one has

$$\Sigma \vdash P(\ulcorner \varphi \urcorner) \text{ iff } \Sigma \vdash \varphi.$$

[N.B. This is exactly what one needs to obtain the independence of Gödel's sentence. If one is only interested in the unprovability of the Gödel sentence, one needs only the right-to-left implication.]

Under this criterion for expressing provability and the earlier cited one for self-expressibility, Kreisel constructed two predicates expressing provability and one sentence apiece expressing its own provability. One was provable and one was not. Let us look at these examples.

EXAMPLE I. There is an expression $P_1(v)$ of provability and a sentence φ expressing its own provability (relative to P_1) such that $\Sigma \vdash \varphi$.

Proof. Once again, the provabilities of

$$\bar{0} = \bar{0} \leftrightarrow Pr(\ulcorner \bar{0} = \bar{0} \urcorner) \quad \text{and} \quad \bar{0} = \bar{0}$$

are *irrelevant* as $\overline{0} = \overline{0}$ is merely equivalent to and does not express its own provability.

Kreisel begins by choosing φ via Gödel's construction to express $Pr(\ulcorner \varphi \to \varphi \urcorner)$. He then takes $P_1(x)$ to be $Pr(\ulcorner \varphi \urcorner \dotrightarrow x)$, where $\dotrightarrow$ denotes the primitive recursive function such that for all sentences

$$\ulcorner \chi_1 \urcorner \dotrightarrow \ulcorner \chi_2 \urcorner = \ulcorner \chi_1 \to \chi_2 \urcorner$$

for all sentences χ_1, χ_2.

φ expresses $P_1(\ulcorner \varphi \urcorner)$ in Kreisel's sense and it remains only to check that $P_1(v)$ expresses provability in Kreisel's sense. In fact, as Kreisel points out, $P_1(v)$ is equivalent to $Pr(v)$ and the latter expresses provability as was known, e.g., from Hilbert and Bernays (1939).

Kreisel merely remarks that $P_1(v)$ and $Pr(v)$ are equivalent. It would be interesting to see how he proved this as the available Hilbert–Bernays Derivability Conditions were rather awkward. I shall derive this equivalence using the more managable Löb Derivability Conditions. Both sets of conditions will be listed in full in the next section; here let me simply cite the two from Löb's list that I need:

D1. $\Sigma \vdash \theta \Rightarrow \Sigma \vdash Pr(\ulcorner \theta \urcorner)$
D2. $\Sigma \vdash Pr(\ulcorner \theta \urcorner) \wedge Pr(\ulcorner \theta \to \chi \urcorner) \to Pr(\ulcorner \chi \urcorner)$,

for any sentences θ, χ.

First, observe: For φ expressing $Pr(\ulcorner \varphi \to \varphi \urcorner)$,

$$\Sigma \vdash \varphi \to \varphi \Rightarrow \Sigma \vdash Pr(\ulcorner \varphi \to \varphi \urcorner), \quad \text{by D1}$$

$$\Rightarrow \Sigma \vdash \varphi, \quad \text{by choice of } \varphi$$

$$\Rightarrow \Sigma \vdash Pr(\ulcorner \varphi \urcorner), \tag{1}$$

by D1. Also,

$$\Sigma \vdash P_1(\ulcorner \psi \urcorner) \to P_1(\ulcorner \psi \urcorner) \wedge Pr(\ulcorner \varphi \urcorner), \quad \text{by (1)}$$

$$\vdash P_1(\ulcorner \psi \urcorner) \to Pr(\ulcorner \varphi \urcorner) \wedge Pr(\ulcorner \varphi \to \psi \urcorner), \quad \text{by choice of } P_1$$

$$\vdash P_1(\ulcorner \psi \urcorner) \to Pr(\ulcorner \psi \urcorner), \quad \text{by D2}.$$

This is half of what we want. For the converse, observe

$$\Sigma \vdash \psi \to (\varphi \to \psi) \Rightarrow \Sigma \vdash Pr(\ulcorner \psi \to (\varphi \to \psi) \urcorner), \quad \text{by D1}$$

$$\Rightarrow \Sigma \vdash Pr(\ulcorner \psi \urcorner) \to Pr(\ulcorner \varphi \to \psi \urcorner), \quad \text{by D2}$$

$$\Rightarrow \Sigma \vdash Pr(\ulcorner \psi \urcorner) \to P_1(\ulcorner \psi \urcorner), \quad \text{by choice of } P_1. \qquad \square$$

In a footnote in his paper, Kreisel gives a simpler example supplied by Henkin. The example cited here is, however, of some interest.

EXAMPLE II. There is an expression P_2 of provability and a sentence ψ expressing its own provability (relative to P_2) and such that $\Sigma \nvdash \psi$.

Here, Kreisel begins by choosing ψ to express

$$[\ulcorner\psi\urcorner \neq \ulcorner\psi\urcorner \to Pr(\ulcorner\psi\urcorner)] \wedge [\ulcorner\psi\urcorner = \ulcorner\psi\urcorner \to Pr(\ulcorner\bar{0} = \bar{1}\urcorner)]$$

and then defines

$$P_2(x) : [x \neq \ulcorner\psi\urcorner \to Pr(x)] \wedge [x = \ulcorner\psi\urcorner \to Pr(\ulcorner\bar{0} = \bar{1}\urcorner)].$$

The details of the proof that this works do not interest us here. Again, in a footnote, Kreisel offers a simpler variant due to Henkin.

It is amusing to note that if φ is any sentence other than ψ satisfying

$$\Sigma \vdash \varphi \leftrightarrow P_2(\ulcorner\varphi\urcorner),$$

then

$$\Sigma \vdash \varphi \leftrightarrow Pr(\ulcorner\varphi\urcorner)$$

and Löb's Theorem (cf. the next section) yields

$$\Sigma \vdash \varphi.$$

If one applies Gödel's construction to P_2, the sentence φ one constructs will *not* be ψ—for ψ is of the form $P_2(t_1)$ where t_1 refers to

$$[x \neq x \to Pr(x)] \wedge [x = x \to Pr(\ulcorner\bar{0} = \bar{1}\urcorner)],$$

while φ is of the form $P_2(t_2)$ where t_2 refers to $P_2(x)$. Hence, Kreisel does not solve the problem if we do not relax the restriction on self-expression to not requiring the actual sentence Gödel's method constructs for P_2.

Incidentally, this example, in which distinct fixed points behave differently, though not atypical of self-reference, is atypical of metamathematical applications of self-reference. However, it would take more experience to recognize this and to relax the demand of self-referential expression to an acceptance of mere provable equivalence.

2. Löb's Solution, Löb's Theorem, and Kreisel's Response

To discuss Löb's solution to Henkin's problem and see how close and yet how far Kreisel's Example I was from it, we must back up a bit and recall Paul Bernays' proof (in *Grundlagen der Mathematik II*) of the Second Incompleteness Theorem. This proof is not so simple as that currently expounded (say, that in my *Self-Reference and Modal Logic*). Simplification comes with experience and Bernays was offering the first detailed proof of the Second Incompleteness Theorem. In his published paper Gödel (1931) had merely remarked that his proof showed

$$\Sigma \vdash \mathrm{Con}(\Sigma) \to \neg Pr(\ulcorner\varphi\urcorner), \tag{1}$$

where φ was his sentence asserting its own unprovability, whence

$$\Sigma \vdash \mathrm{Con}(\Sigma) \to \varphi. \tag{2}$$

Thus, since φ is unprovable, (2) shows that $\mathrm{Con}(\Sigma)$ is unprovable. The full details of the derivation of (1) were left by Gödel to a sequel he never published, i.e., he left the task of expounding them to others. Bernays performed this task. Basically, Bernays proved the Second Incompleteness Theorem by formalizing the proof of the First Incompleteness Theorem, using as lemmas to handle some of the details certain *derivability conditions*, now known as the Hilbert–Bernays Derivability Conditions:

HB1. $\Sigma \vdash \varphi \to \psi \Rightarrow \Sigma \vdash Pr(\ulcorner\varphi\urcorner) \to Pr(\ulcorner\psi\urcorner)$.

HB2. $\Sigma \vdash Pr(\ulcorner\neg\varphi v\urcorner) \to Pr(\ulcorner\neg\varphi\bar{n}\urcorner)$, for each numeral $\bar{n}$.

HB3. If f is a primitive recursive unary function and $\bar{f}$ is a function constant denoting it,

$$\Sigma \vdash \bar{f}v = \bar{0} \to Pr(\ulcorner\bar{f}v = \bar{0}\urcorner),$$

where $\ulcorner\varphi\dot{v}\urcorner$ indicates, not $\ulcorner\varphi v\urcorner$, but the more complicated sub($\ulcorner\varphi v\urcorner$, $num(v)$), the term with free variable v denoting the Gödel number of the result of replacing v in φv by the vth numeral.

The Hilbert–Bernays Derivability Conditions, which came out in 1939, were the closest thing to an attempt to analyze $Pr(v)$ before Kreisel replaced them by the criterion of expressibility cited in the previous section:

CE. $\Sigma \vdash P(\ulcorner\varphi\urcorner)$ iff $\Sigma \vdash \varphi$.

For his predicates P, Kreisel had only CE to use, although he evidently used more about $Pr(v)$, but he never codified what he used. The Hilbert–Bernays Derivability Conditions codify a bit more, but they were technical lemmas in a single proof and clearly were not intended as an analysis of $Pr(v)$: HB2, for example, has the obvious improvement:

$$\Sigma \vdash Pr(\ulcorner\varphi v\urcorner) \to Pr(\ulcorner\varphi t\urcorner), \quad t \text{ any term}.$$

Moreover, the extension of HB3 to functions of several variables is also obvious.

In 1954, Martin Hugo Löb doubly announced his solution to Henkin's problem. The lesser of the two announcements was a contributed talk at the International Congress of Mathematicians in Amsterdam, the proceedings of which conference includes his abstract (Löb, 1954) for this talk. The greater of the two announcements was a paper (Löb, 1955) submitted to the *Journal of Symbolic Logic*, the paper, augmented by comments of the referee, appeared the following year. Löb's paper is notable for four things:

(1) the introduction of the new Löb Derivability Conditions (or, *the* Derivability Conditions);
(2) the solution of Henkin's problem for sentences *expressing* their own provabilities;
(3) the solution of the more general problem in which sentences are merely required to be equivalent to their own provabilities;
(4) a further generalization called Löb's Theorem.

The ICM abstract contains (1) and (2); contributions (3) and (4) occur only in the JSL paper and Löb credits the anonymous referee (Henkin) for noticing that his proof of (2) actually established (3) and (4). Contribution (2) was, of course, the heart of the matter; nonetheless, contributions (1) and (4) are the most important.

Löb's Derivability Conditions (in my preferred order) are:

D1. $\Sigma \vdash \varphi \Rightarrow \Sigma \vdash Pr(\ulcorner\varphi\urcorner)$,
D2. $\Sigma \vdash Pr(\ulcorner\varphi\urcorner) \wedge Pr(\ulcorner\varphi \rightarrow \psi\urcorner) \rightarrow Pr(\ulcorner\psi\urcorner)$,
D3. $\Sigma \vdash Pr(\ulcorner\varphi\urcorner) \rightarrow Pr(\ulcorner Pr(\ulcorner\varphi\urcorner)\urcorner)$,

for all sentences φ, ψ. In his paper, he actually lists two more, namely HB1 and HB3. However, HB1 is redundant, HB3 is used only to derive D3, and only D1–D3 are used in the subsequent proof, and it has thus become customary to list only these three conditions. As Löb pointed out, D1 can be found in Hilbert–Bernays although Bernays did not list it among his "Ableitbarkeitsforderungen" HB1–HB3. Moreover, D2 is an improved version of HB1, and, as just mentioned, D3 is a consequence of HB3. Thus, Löb's Derivability Conditions do not in any way constitute a technical break-through. They are marvellous nonetheless, for they constitute the beginning of an analysis of $Pr(v)$, a step toward the proper generality, and useful lemmas —as we have already seen in discussing Kreisel's first example and as we shall see in discussing Löb's Theorem.

Löb's Theorem, isolated by Henkin as the general result established by Löb's proof once one removes the unnecessary premises is the following:

Löb's Theorem. *Let ψ be any sentence. If $\Sigma \vdash Pr(\ulcorner\psi\urcorner) \rightarrow \psi$, then $\Sigma \vdash \psi$.*

Proof. Assume

$$\Sigma \vdash Pr(\ulcorner\psi\urcorner) \rightarrow \psi. \tag{1}$$

Let φ be such that

$$\Sigma \vdash \varphi \leftrightarrow . Pr(\ulcorner\varphi\urcorner) \rightarrow \psi. \tag{2}$$

(Actually, Löb constructs φ expressing $Pr(\ulcorner\varphi\urcorner) \rightarrow \psi$. Even though the theorem is stated in the mathematically most natural terms, in his proof Löb still clings to the philosophical formulation.) Observe

$$\Sigma \vdash Pr(\ulcorner\varphi\urcorner) \leftrightarrow Pr(\ulcorner Pr(\ulcorner\varphi\urcorner) \rightarrow \psi\urcorner), \quad \text{by (2), D1, D2}$$

$$\vdash Pr(\ulcorner\varphi\urcorner) \rightarrow Pr(\ulcorner Pr(\ulcorner\varphi\urcorner)\urcorner) \rightarrow Pr(\ulcorner\psi\urcorner), \quad \text{by D2}$$

$$\vdash Pr(\ulcorner\varphi\urcorner) \rightarrow Pr(\ulcorner\psi\urcorner), \quad \text{by D3}$$

$$\vdash Pr(\ulcorner\varphi\urcorner) \rightarrow \psi, \tag{3}$$

by assumption (1). But (2) and (3) yield $\Sigma \vdash \varphi$, whence

$$\Sigma \vdash Pr(\ulcorner\varphi\urcorner), \quad \text{by D1}$$

$$\vdash \psi, \quad \text{by (3)}. \qquad \square$$

There is so much to say about this proof I scarcely know where to begin. It is such a delight that one would expect it to have led to a lot of similar playing around. Yet it didn't Why? The explanation I hinted at in my review of Boolos' book (Smoryński, 1981a) may strike one as facetious and is definitely quite naïve, but there may be something to it: Löb's Theorem is of supreme metamathematical (hence: philosophical) importance: Where Gödel's (philosophically important) Second Incompleteness Theorem merely asserts the consistency of a theory to be unprovable in the theory, Löb's Theorem transcends this and characterizes those instances of soundness provable in the theory as those trivially so provable. My explanation was that Löb's Theorem became sacrosanct and nobody dared to "play around" with it.

In any event, Löb's paper tied Löb's Theorem a second time to philosophy by remarking that the referee (the ubiquitous Henkin) had uncovered a new paradox underlying the fixed point $\varphi \leftrightarrow Pr(\ulcorner\varphi\urcorner) \to \psi$: Imagine a sentence asserting

"If I am true, then so is A."

If B is this sentence,

$$B = B \to A,$$

then B can only be false if B is true and A is false. Thus B is true. But then $B \to A$ is true, whence A is true—and this, regardless of what A is. (It turns our that this paradox was not brand new: It pops up in discussing Haskell Curry's paradoxical combinator in Combinatory Logic—cf. van Bentham (1978) for discussion.)

A somewhat less emotional hypothesis than the one that philosophy frightens off mathematicians is that philosophy guides one's actions and that again the wrong philosophy was in place. Perhaps the discovery of the new paradox showed that the old paradigms were still in place and mathematical experimentation was not called for. Thus, for example, it was not noticed until the 1970s that the solution to the generalized Henkin problem,

$$\Sigma \vdash \varphi \leftrightarrow Pr(\ulcorner\varphi\urcorner) \Rightarrow \Sigma \vdash \varphi, \tag{*}$$

was a general as Löb's Theorem,

$$\Sigma \vdash Pr(\ulcorner\psi\urcorner) \to \psi \Rightarrow \Sigma \vdash \psi.$$

For, if

$$\Sigma \vdash Pr(\ulcorner\psi\urcorner) \to \psi,$$

then $\varphi = \psi \wedge Pr(\ulcorner\psi\urcorner)$ satisfies

$$\Sigma \vdash \varphi \leftrightarrow Pr(\ulcorner\varphi\urcorner),$$

and (*) yields

$$\Sigma \vdash \varphi$$

$$\vdash \psi.$$

The only one who (to my knowledge) can be said to have played around with this stuff prior to the 1970s is Kreisel, who in various expositions (e.g., Kreisel (1968) and Kreisel and Takeuti (1974)) derived Löb's Theorem using the alternate fixed point

$$\Sigma \vdash \varphi \leftrightarrow Pr(\ulcorner \varphi \to \psi \urcorner),$$

where it is assumed that

$$\Sigma \vdash Pr(\ulcorner \psi \urcorner) \to \psi.$$

Kreisel appears to have been quite pleased with this fixed point and asserted in his joint paper with Takeuti that the proof using his fixed point is more general than Löb's in that it only uses positive intuitionistic reasoning. As Löb's proof clearly assumes no more (but cf. Section 4, below), one must assume his preference really to be based on something else. Personally, I cannot help but noticing a similarity between his new fixed point

$$\varphi \leftrightarrow Pr(\ulcorner \varphi \to \psi \urcorner) \tag{$*$}$$

and his old predicate

$$P_1(\ulcorner \psi \urcorner) = Pr(\ulcorner \varphi \to \psi \urcorner).$$

In this latter, φ was chosen by

$$\varphi \leftrightarrow Pr(\ulcorner \varphi \to \varphi \urcorner).$$

Choosing, instead, φ to depend on ψ, one just gets ($*$). In any event, the superiority of one of Kreisel's or Löb's fixed points over the other seems, at this conceptual level, to boil down to: Löb's fixed point yields a marginally shorter derivation of Löb's Theorem and Kreisel's yields a marginally shorter proof of the Formalized Löb's Theorem:

Formalized Löb's Theorem. *For any sentence* ψ,

$$\Sigma \vdash Pr(\ulcorner Pr(\ulcorner \psi \urcorner) \to \psi \urcorner) \to Pr(\ulcorner \psi \urcorner).$$

Proof. Choose φ so that

$$\Sigma \vdash \varphi \leftrightarrow Pr(\ulcorner \varphi \to \psi \urcorner).$$

From this equivalence, the instance,

$$\Sigma \vdash Pr(\ulcorner \varphi \to \psi \urcorner) \to Pr(\ulcorner Pr(\ulcorner \varphi \to \psi \urcorner) \urcorner),$$

of D3, and application of D1, D2, we get

$$\Sigma \vdash \varphi \to Pr(\ulcorner \varphi \urcorner)$$

$$\vdash \varphi \to Pr(\ulcorner \varphi \urcorner) \wedge Pr(\ulcorner \varphi \to \psi \urcorner), \quad \text{by choice of } \varphi$$

$$\vdash \varphi \to Pr(\ulcorner \psi \urcorner).$$

Conversely,

$$\Sigma \vdash Pr(\ulcorner \psi \urcorner) \to Pr(\ulcorner \varphi \to \psi \urcorner), \quad \text{by D1, D2}$$

$$\vdash Pr(\ulcorner \psi \urcorner) \to \varphi, \quad \text{by choice of } \varphi.$$

Thus, $\Sigma \vdash \varphi \leftrightarrow Pr(\ulcorner \psi \urcorner)$ and D1, D2 again yield

$$\Sigma \vdash Pr(\ulcorner \psi \urcorner) \leftrightarrow Pr(\ulcorner Pr(\ulcorner \psi \urcorner) \rightarrow \psi \urcorner),$$

which is slightly more than desired. □

3. The Modern Period, I: Solovay's Theorem

The proof of the Formalized Löb's Theorem given at the end of the last section is in the more modern, *abstract* spirit. There is no reference to the construction either of φ or of $Pr(v)$. The proof uses the existence of φ merely as something given by an existence proof as something satisfying the property

$$\Sigma \vdash \varphi \leftrightarrow Pr(\ulcorner \varphi \urcorner);$$

it uses Löb's Derivability Conditions as established lemmas in an almost axiomatic framework. This is what has happened to the subject: In the 1970s, this axiomatic approach, divorced from philosophy (in particular, from the paradoxes), developed. Moreover, it developed among a new, younger set of researchers. Before encountering this development, however, I want to take a parting glance at Kreisel.

For years, Kreisel appears to have been the only one interested in formal linguistic self-reference. He is the only one who consistently wrote on the subject. Still, we can trace two common themes in his writings back to his 1953 paper. One theme is the question of what it means to express provability. In an article in *Lectures on Modern Mathematics III* (Kreisel, 1965), he discussed a notion of *canonicality* of proof predicates in attempting to isolate $Pr(v)$ from other candidates $P(v)$. Ultimately, he decided that $Pr(v)$ is canonical for the usual *presentation* of Σ, but that his predicates $P(v)$ are also canonical for oddly presented formal systems, that Rosser's Theorem is not based on "Rosser's trick" but on a canonical Rosser proof predicate for another oddly presented formal system, and that there is a similar canonical proof predicate for naturally, if untraditionally, presented formal systems such as cut-free analysis. Moreover, for these various canonical (whence: consideration-worthy) predicates, it may still be the case that there is a difference between sentences expressing certain properties of themselves and mere fixed points. His joint paper with Gaisi Takeuti, published in 1974, is the clearest example of this: For the usual systems Σ, any Gödel sentence (i.e., any sentence φ such that $\Sigma \vdash \varphi \leftrightarrow \neg Pr(\ulcorner \varphi \urcorner)$) is equivalent to $\mathrm{Con}(\Sigma)$, whence any two Gödel sentences are provably equivalent. And, any two Henkin sentences (sentences φ such that $\Sigma \vdash \varphi \leftrightarrow Pr(\ulcorner \varphi \urcorner)$) are provably equivalent (since they are provable). Kreisel and Takeuti investigate these sentences and the Rosser sentence (not defined here) for the special system Σ of Cut-Free Analysis, i.e., a second-order theory of arithmetic based on direct rules of inference. They prove that:

(i) any two Gödel sentences for this special Σ are provably equivalent;
(ii) each literal Henkin sentence (i.e., each sentence expressing its own prov-

ability) is refutable in Σ, but that there are provable Henkin sentences (e.g., $\overline{0} = \overline{0}$); and

(iii) all literal Rosser sentences for Σ are equivalent, but not all Rosser sentences are.

Mathematically, this joint work with Takeuti is much deeper than anything in self-reference that had gone before, but this depth is in the proof theory of cut-free analysis. In the introduction to the paper, Kreisel suggests that those who have "doubts about the significance" of cut-free analysis can view the paper as supplying "'counterexamples' in a general theory of self-referential propositions of formal systems." As work in self-reference, this is exactly the perspective by which to view the paper: It offers no positive theory, it fails to give direction for further research (self-referentially, it has no sequel), and its focus is extremely narrow. It is safe to say that Kreisel (the author principally responsible for the self-referential material) has gone beyond the paradigm of the paradoxes, and one might say he is in that transitional period in which problems arise bringing forth the new paradigms. In any event, with this paper, he is not in the modern (i.e., current) period whereby the uniqueness and explicit definability of the Gödel and Henkin sentences have been nicely generalized and the foundations of a limited "general theory of self-referential propositions," against which his counterexamples can be placed, have been laid. To reach the current period, a conceptual shift is necessary.

The conceptual shift needed—if only in the accidental, historical sense—is away from intension: The notion of a sentence's expressing something about itself has not proven fruitful, nor has the notion of canonicality of expression of a proof predicate, nor has the insight that a Rosser sentence is a Gödel sentence for a Rosser proof predicate. What has proven fruitful is a more axiomatic approach: One thinks of *any* proof predicate and works with the properties it has (e.g., D1–D3). Ultimately, this yields an analysis of methods; crudely, it leads to modal logic, the new predicate being treated like a (black) box accessible only by some list of its properties. The chief results of this modal analysis have become the new paradigms governing research in formal linguistic self-reference. These results were, however, not the real breakthroughs— the gestalt shifts; these results were possible because a conceptual change had already taken place.

The conceptual change cannot chronologically be pinpointed with any accuracy. There were several modern beginnings and a few precursors, such as Montague's (1963) paper. As I do not view the assignation of credit to be the task of the historian, I shall simply choose the first published paper nearing the modern perspective as a typical enough beginning.

Angus Macintyre and Harry Simmons are mathematical logicians, not philosophical ones. They did not like self-reference in arithmetic. As Simmons once confided in me, they could not understand how their friend Donald Jensen was able to concoct self-referential formulas which would allow him to prove various lemmas. In an article Macintyre and Simmons (1973) set about on the task of replacing diagonalization by some one major consequence thereof. The strongest consequence they could think of was Löb's Theorem.

At the time their paper came out, I recall being of two minds. I liked their paper, but I could not find anything about it that I could put my finger on and say "This is good; this is what I like." Kreisel's review pinpoints the problem: "They disregard the familiar asymmetry between positive and negative metamathematical results, and thereby produce a host of teratological problems." The Macintyre–Simmons paper is a transitional one in which a new problem area is being introduced and some concepts are being provisionally put forward, but they have not reached their final forms. It was only after these concepts crystallized that one could point to this paper and say "Macintyre and Simmons accomplished x, y, and z."

In a serious, thorough-going history of self-reference one would have to go into this paper carefully and consider the other origins of the modern subject. (I know of three: Löb and de Jongh in Amsterdam, Magari and Sambin in Italy, and Boolos and Solovay in the United States. There are others who may have been close.) Here, I shall simply state the new framework, then go back to say what was accomplished by Löb, Macintyre and Simmons, and others, and finally I shall say a little about the new paradigms that ultimately arose in this framework. The new framework is, as I have already noted, modal logic. The idea is to treat $Pr(v)$ as necessity and use the language and tools of modal logic to study $Pr(v)$.

I assume the reader is familiar with the notation of modal logic. (If he is not, the best place I can refer him to for the following discussion is the very reasonably priced *Self-Reference and Modal Logic* (Smoryński, 1985).) If we let the modal box, $\Box$, stand for provability, what are the obvious properties of the box? The Derivability Conditions yield the axioms and rules of inference:

D1. $A / \Box A$;
D2. $\Box A \land \Box(A \to B) \to \Box B$;
D3. $\Box A \to \Box \Box A$.

Diagonalization can be modally formulated, either by the addition of special constants or as a rule of inference (cf. Chapter 1 of Smoryński (1985) for details); but, if like Macintyre and Simmons we want to avoid diagonalization, the strongest known result was Löb's Theorem,

LT. $\Box A \to A / A$,

or the Formalized Löb's Theorem,

FLT. $\Box(\Box A \to A) \to \Box A$.

The modal language gives us a great deal. No mean prize it affords us is a simple criterion of elegance by means of which we can objectively declare Löb's Derivability Conditions to be more elegant than Bernays'. The extra-modal-propositional character of Löb's reference to the sentence φ expressing $Pr(\ulcorner\varphi\urcorner) \to \psi$ is now not merely inelegant, but it has also become extraneous to the point that it will puzzle the modern reader—provided he notices that Löb is not merely repeating the construction of φ for the sake of exposition. The modal language also explains clearly what I had only hinted at in the beginning of this section in stating how pleasant the proof at the end of Section 2 of the Formalized Löb's Theorem was: The proof is a purely modal propositional deduction.

What Macintyre and Simmons accomplished, put in these modal terms, was a number of things. Working with D1–D3 as the modal axioms and rule of inference of a basic modal system, which I shall call BML in accordance with Smoryński (1985), they showed:

(1) The schema LT is equivalent to the schema FLT over BML.
(2) The schema LT is equivalent over BML to the existence of fixed points p to $\Box p \to A$ and, moreover, such p is equivalent to $\Box A \to A$.
(3) A similar equivalence over BML obtains between FLT and the existence of (definable) fixed points p to $\Box(p \to A)$.

On an absolute scale, Macintyre and Simmons cannot be said to have accomplished much; on a relative scale, ignoring the disparity of philosophical interest and ignoring any considerations of amount of effort, we find more here than in Löb's paper, more than in Kreisel's proof of Löb's Theorem, even more than in Gödel's paper! The Macintyre–Simmons paper is a watershed, but a skewed one—it lies more heavily on *this* (i.e., the modern) side of the watershed: It is the first paper on self-reference with so much in it and yet which is not universally recognized as a milestone, which is "just another paper" as it were. It is the last paper that needs to be mentioned because of its mere existence; henceforth, all references are selective.

I have said before that writing history is not sorting out credit. Nor is it preparing a chronology. It is, rather, an explanation of a development. Following Macintyre and Simmons, the rest of my story is inevitable: The problems were "in the air," there was a great deal of talent out there, ... The net result was the following three theorems, which have become the new paradigms.

Uniqueness Theorem. *Let* PrL *be the extension of* BML *by the addition of* FLT. *Let, further,* $A(p)$ *be a formula in which every occurrence of* p *lies inside the scope of a* $\Box$. *Then*:

$$\text{PrL} \vdash \Box(p \leftrightarrow A(p)) \wedge \Box(q \leftrightarrow A(q)) \to \Box(p \leftrightarrow q).$$

Explicit Definability Theorem. *Let* p, $A(p)$ *be as in the preceding result. There is a modal formula* D *possessing only those atoms in* A *other than* p *and such that*

$$\text{PrL} \vdash D \leftrightarrow A(D).$$

Completeness Theorem. *For any modal A,*

$$\text{PrL} \vdash A \text{ iff } \Sigma \vdash A^*,$$

*for every interpretation * in Σ in which $\square$ is interpreted by $Pr(v)$.*

The first of these generalizes the uniqueness of the Gödel and Henkin sentences and, following Macintyre and Simmons, that of the Löb and Kreisel sentences. It seems to be the "proper" generalization. The Uniqueness Theorem is not particularly difficult: it was independently proven by Claudio Bernardi (1976), Dick de Jongh (unpublished, cf. Chapter 2 of Smoryński (1985), and Giovanni Sambin (1976) by three different methods. Bernardi's proof generalizes most easily and will be given in this yet greater generality in the next section.

The Explicit Definability Theorem is due to de Jongh (unpublished, cf. Smoryński (1979, 1985)) and Sambin (1976) following partial results of Bernardi (1975) and Smoryński (1979). The ultimate proofs of de Jongh and Sambin are rather more similar than different and again generalize nicely. There are now other proofs, but these do not generalize (or, better: have not done so yet). The full proof (in generality) can be found in Smoryński (1985); I shall not present it here as it would take too much space. However, I do note that it is an induction in which Löb's Theorem is applied during the basis step.

The Completeness Theorem is most interesting. It is the deepest of the three results, a fact testified to by knowledge of the proofs as well as such external indicators as the fact that it was proven solely by Robert M. Solovay even though it was known as an open problem by all the other researchers. In richness of consequence, as well as depth of proof, Solovay's Theorem is again the deepest of the three results—cf. Chapter 3 of Smoryński (1985) for details.

Solovay's Theorem is more of meta-historic importance to us. It offers a mathematical explanation of why, for example, Macintyre and Simmons could not find anything stronger than Löb's Theorem in their attempt to replace diagonalization: In the restricted modal language under consideration, there is nothing stronger. In a sense, Solovay's Theorem completes a chapter in the history of self-reference.

4. The Modern Period, II: A Generalization of Löb's Theorem

As one chapter ends, another usually begins. Modern work on self-reference in arithmetic has primarily been in pursuit of analogues to the three paradigms. Franco Montagna has, for example, investigated the "Feferman predicate" of Feferman (1960) in one paper (Montagna, 1978) and the broader context of quantified modal logic for the familiar predicate $Pr(v)$ in another paper (Montagna, 1984). In the latter paper, for example, he showed that the

Uniqueness Theorem generalized, and that the Explicit Definability Theorem failed modally, but so did Completeness (relative to the obvious axioms). In a sequel, Smoryński (1987) completed the second task by showing that the Explicit Definability Theorem also failed arithmetically.

The new programmes can be described thus: Let $\rho(v)$ be a new predicate and consider either a modal logic with $\square$ denoting ρ or with a new operator ∇ denoting ρ and $\square$ still denoting Pr; or generalize the language in some other way. Determine whether or not fixed points are unique in this language. Find explicit definitions of the fixed points where possible. Find a complete axiomatization.

The difference between, say, the counterexamples of Montagna and Smoryński on the one hand and those of Kreisel and Takeuti on the other is this new set of paradigms. The new counterexamples are derived with an eye to the general theory and are stated less anecdotally. (Of course, old counterexamples can be looked at anew.) Another difference is that the paradigms are *general* positive results to aim at and greater success may now be possible—as one sees by contrasting Montagna's generalized Uniqueness Theorem to Kreisel and Takeuti's isolated uniqueness result for Gödel sentences.

Because of their close connection with Löb's Theorem, I wish in this section to discuss the currently strongest generalizations of the Uniqueness and Explicit Definability Theorems. These are my own results, but they merit discussion nonetheless and, moreover, they are not so deep that I will be in danger of appearing to show off.

The Uniqueness Theorem tells us that a lot of self-referential sentences are unique (up to provable equivalence): If $\rho(v)$ is an arithmetical instance of a modal formula, and if

$$\Sigma \vdash \varphi_1 \leftrightarrow \rho(\ulcorner\varphi_1\urcorner), \qquad \Sigma \vdash \varphi_2 \leftrightarrow \rho(\ulcorner\varphi_2\urcorner),$$

then

$$\Sigma \vdash \varphi_1 \leftrightarrow \varphi_2.$$

How many more ρ's does this hold for? It turns out that a simple property of ρ is all that is needed to apply Bernardi's proof (slightly simplified!) to obtain the uniqueness of the fixed point for ρ.

Definition. A formula $\rho(v)$, with only v free, is *substitutable* iff, for all sentences φ, ψ,

$$\Sigma \vdash Pr(\ulcorner\varphi \leftrightarrow \psi\urcorner) \rightarrow \rho(\ulcorner\varphi\urcorner) \leftrightarrow \rho(\ulcorner\psi\urcorner).$$

Two weaker conditions worth mentioning in passing are extensionality and provable extensionality.

Definitions. A formula $\rho(v)$, with only v free, is *extensional* iff, for all sentences φ, ψ,

$$\Sigma \vdash \varphi \leftrightarrow \psi \Rightarrow \Sigma \vdash \rho(\ulcorner\varphi\urcorner) \leftrightarrow \rho(\ulcorner\psi\urcorner).$$

ρ is *provably extensional* iff, for all sentences φ, ψ,

$$\Sigma \vdash Pr(\ulcorner\varphi \leftrightarrow \psi\urcorner) \to Pr(\ulcorner\rho(\ulcorner\varphi\urcorner) \leftrightarrow \rho(\ulcorner\psi\urcorner)\urcorner).$$

For a Σ_1-sound theory Σ, provable extensionality implies extensionality. Moreover, by D3, substitutability implies provable extensionality. We will prove in a moment that the Uniqueness Theorem holds if ρ is substitutable. It does not hold for general extensional or provably extensional formulas. For example, Albert Visser has observed that Feferman's predicate is provably extensional yet has inequivalent fixed points. (Here, one might ask, à la Kreisel and Takeuti (or, indeed: à la Henkin), about literal fixed points, i.e., are sentences *expressing* their provabilities with respect to Feferman's predicate provably equivalent?)

Generalized Uniqueness Theorem. *Let $\rho(v)$ be substitutable. Let φ, ψ be sentences satisfying*

$$\Sigma \vdash \varphi \leftrightarrow \rho(\ulcorner\varphi\urcorner), \qquad \Sigma \vdash \psi \leftrightarrow \rho(\ulcorner\psi\urcorner).$$

Then $\Sigma \vdash \varphi \leftrightarrow \psi$.

Proof. This is a straightforward adaptation of Bernardi's proof of the Uniqueness Theorem for $A(p)$ in PrL. To illustrate the modal-propositional nature of the proof, I will carry out the proof in a modal system I call SR⁻ (from *self-reference*). SR⁻ is obtained by adding to PrL a new modal operator ∇ and an axiom schema asserting its substitutability,

S. $\Box(A \leftrightarrow B) \to . \nabla A \leftrightarrow \nabla B$.

We want to show:

$$\text{SR}^- \vdash \Box(p \leftrightarrow \nabla p) \wedge \Box(q \leftrightarrow \nabla q) \to \Box(p \leftrightarrow q).$$

Observe,

$$\text{SR}^- \vdash \Box(p \leftrightarrow q) \to . \nabla p \leftrightarrow \nabla q, \quad \text{by } S$$

$$\vdash \quad \Box\,\Box(p \leftrightarrow q) \to \Box(\nabla p \leftrightarrow \nabla q)$$

$$\vdash \quad \Box(p \leftrightarrow \nabla p) \wedge \Box(q \leftrightarrow \nabla q)$$

$$\to [\Box\,\Box(p \leftrightarrow q) \to \Box(p \leftrightarrow q)] \qquad\qquad (*)$$

$$\vdash \quad \Box(p \leftrightarrow \nabla p) \wedge \Box(q \leftrightarrow \nabla q)$$

$$\to \Box[\Box\,\Box(p \leftrightarrow q) \to \Box(p \leftrightarrow q)]$$

$$\vdash \quad \Box(p \leftrightarrow \nabla p) \wedge \Box(q \leftrightarrow \nabla q) \to \Box\,\Box(p \leftrightarrow q),$$

by FLT, whence

$$\text{SR}^- \vdash \Box(p \leftrightarrow \nabla p) \wedge \Box(q \leftrightarrow \nabla q) \to \Box(p \leftrightarrow q), \quad \text{by } (*),$$

where the unexplained steps are derived by applications of D1–D3. $\qquad\Box$

One can similarly derive a generalization to fixed points to $A(p)$ for p lying only in the scopes of $\square$'s and ∇'s in A. The crucial point is the substitutability of $A(p)$, which follows from the appropriate modal substitution lemma for SR^-.

As for the question of the Explicit Definability of fixed points, I have shown (Smoryński, 1987) that the substitutability of ρ is not enough to guarantee the explicit definability (in the appropriate language) of any fixed point

$$\Sigma \vdash \varphi \leftrightarrow \rho(\ulcorner\varphi\urcorner).$$

However, if one assumes ρ to be Σ_1, the Explicit Definability Theorem holds for all appropriate formulas constructed from $Pr(v)$ and $\rho(v)$. In modal terms, one extends SR^- to a modal theory SR by an axiom asserting ρ to have the one stable property of Σ_1-formulas: demonstrable completeness,

DC. $\nabla A \to \square \nabla A.$

Then one can prove:

Generalized Explicit Definability Theorem. *Let $A(p)$ be a formula of the language of SR in which every occurrence of p lies in the scope of a modal operator $\square$ or ∇. There is a sentence D possessing only those atoms of A other than p and such that*

$$SR \vdash D \leftrightarrow A(D).$$

A proof of this can be found in Chapter 4 of Smoryński (1985). The key to the proof is a generalization to ∇ of an extension for $\square$ of Löb's Theorem due to Sambin (1974). The proof of this generalization is an easy adaptation of Sambin's proof, which is a nice modal application of FLT. I am tempted to give this proof. Instead, let us look at the following, which can be used as a lemma in an alternate proof of the Generalized Explicit Definability Theorem:

Generalized Formalized Löb's Theorem. *For any sentence A of the language of SR,*

$$SR \vdash \nabla(\nabla A \to A) \leftrightarrow \nabla A.$$

Modally, this is proved as a second application of the just cited generalization of Sambin's result. As I said, I shall not give this latter proof, whence I shall not give the modal proof of the Generalized Formal Löb's Theorem. Instead, for reasons that will be clear, I will give an arithmetic proof of the arithmetic formulation of this theorem.

Proof of the Arithmetic Version. Let $\rho(v)$ be a substitutable Σ_1-formula, let ψ be given and define φ so that

$$\Sigma \vdash \varphi \leftrightarrow \rho(\ulcorner\varphi \to \psi\urcorner). \tag{1}$$

φ is an analogue to Kreisel's fixed point. Given my earlier remark that Kreisel's fixed point is marginally better for proving the Formalized Löb's Theorem and given that we wish to prove a generalization of the Formalized Theorem,

$$\Sigma \vdash \rho(\ulcorner\rho(\ulcorner\psi\urcorner) \to \psi\urcorner) \leftrightarrow \rho(\ulcorner\psi\urcorner),$$

(the unformalized version,

$$\Sigma \vdash \rho(\ulcorner\psi\urcorner) \to \psi \Rightarrow \Sigma \vdash \rho(\ulcorner\psi\urcorner),$$

is not valid in this generality), the choice of fixed point is obvious. (Indeed, I have been unable to obtain the result using an analogue to Löb's fixed point. Perhaps Kreisel's remark, cited in Section 2 above, that his fixed point gives a more general proof than Löb's has something to it after all.)

We will prove the theorem by showing

$$\Sigma \vdash \varphi \leftrightarrow \rho(\ulcorner\psi\urcorner). \tag{2}$$

For, then one has

$$\Sigma \vdash \varphi \to \psi . \leftrightarrow . \rho(\ulcorner\psi\urcorner) \to \psi$$
$$\vdash Pr(\ulcorner\varphi \to \psi . \leftrightarrow . \rho(\ulcorner\psi\urcorner) \to \psi\urcorner), \quad \text{by D1}$$
$$\vdash \varphi \leftrightarrow \rho(\ulcorner\rho(\ulcorner\psi\urcorner) \to \psi\urcorner), \quad \text{by substitutability}$$
$$\vdash \rho(\ulcorner\psi\urcorner) \leftrightarrow \rho(\ulcorner\rho(\ulcorner\psi\urcorner) \to \psi\urcorner), \quad \text{by (2)}.$$

Since φ is Σ_1,

$$\Sigma \vdash \varphi \to Pr(\ulcorner\varphi\urcorner). \tag{3}$$

But D1–D2 yield

$$\Sigma \vdash Pr(\ulcorner\varphi\urcorner) \to Pr(\ulcorner(\varphi \to \psi) \leftrightarrow \psi\urcorner)$$
$$\vdash Pr(\ulcorner\varphi\urcorner) \to . \rho(\ulcorner\varphi \to \psi\urcorner) \leftrightarrow \rho(\ulcorner\psi\urcorner)$$
$$\vdash Pr(\ulcorner\varphi\urcorner) \to . \varphi \leftrightarrow \rho(\ulcorner\psi\urcorner), \quad \text{by (1)} \tag{4}$$
$$\vdash \varphi \to . \varphi \leftrightarrow \rho(\ulcorner\psi\urcorner), \quad \text{by (3)}$$
$$\vdash \varphi \to \rho(\ulcorner\psi\urcorner).$$

Conversely,

$$\Sigma \vdash \rho(\ulcorner\psi\urcorner) \to Pr(\ulcorner\rho(\ulcorner\psi\urcorner)\urcorner), \quad \text{since } \rho \text{ is } \Sigma_1$$
$$\vdash Pr(\ulcorner\rho(\ulcorner\psi\urcorner) \to \varphi\urcorner) \wedge \rho(\ulcorner\psi\urcorner) \to Pr(\ulcorner\varphi\urcorner), \quad \text{by D2}$$
$$\vdash Pr(\ulcorner\rho(\ulcorner\psi\urcorner) \to \varphi\urcorner) \wedge \rho(\ulcorner\psi\urcorner) \to [\varphi \leftrightarrow \rho(\ulcorner\psi\urcorner)],$$

by (4),

$$\vdash Pr(\ulcorner\rho(\ulcorner\psi\urcorner) \to \varphi\urcorner) \to . \rho(\ulcorner\psi\urcorner) \to \varphi$$
$$\vdash \rho(\ulcorner\psi\urcorner) \to \varphi, \quad \text{by Löb's Theorem.} \qquad \square$$

As for the third paradigm—a completeness theorem, nothing should be expected for SR⁻ or SR: These modal systems are attempts to analyze specific proofs, not predicates. However, one can specialize the interpretation of ∇ to a fixed ρ, ask what additional axioms are needed, and attempt a completeness theorem for the resulting system. The nicest positive result in this direction is Carlson's Theorem (Carlson, 1985). Other examples can be found in Chapter 4 of Smoryński (1985).

5. Concluding Remarks

Solovay's Completeness Theorem could not have been proven in the 1950s. Its proof, which I have not discussed here, uses Kripke models for modal logic, which models first came on the scene the following decade and spent another decade or so being developed. The other work reported on, however, does not mathematically go beyond the mid-1950s. Except for the notion of a Σ_1-formula, a concept which crystallized somewhat later, and which could even have been replaced by the (more general) proof-generated concept of a demonstrably complete formula, no new notions were needed. What was missing was the right choice of concepts to emphasize, along with a more playful spirit which would have led more quickly to the observations of Macintyre and Simmons, thence, perhaps, to the two new paradigms of Uniqueness and Explicit Definability of *many* fixed points.

I do not wish to overstate the case. These new paradigms are not the only directions to be followed in studying self-reference; they (and one other, unrelated to Löb's Theorem) are merely the currently most successful for our goal of understanding linguistic self-reference—even Solovay's Theorem, which at first sight is not clearly related to self-reference *per se*. There are other developments that seem promising; however, these are not related to Löb's Theorem to which I wish in this paper to narrow our attention.

There are also some mini-developments related to Löb's Theorem that may merit consideration. Foremost among these is a "new" proof of Löb's Theorem which first became well known in the latter half of the 1970s, but which had been known for several years by a number of people. The earliest discovery of it that I know of was by Saul Kripke who hit upon it in 1967 and showed it to a number of people at the UCLA Set Theory Meeting that year. It is very pleasant: Suppose by way of contraposition that $\Sigma \nvdash \psi$. Then $\Sigma + \neg\psi$ is consistent and, by Gödel's Second Incompleteness Theorem,

$$\Sigma + \neg\psi \nvdash \mathrm{Con}(\Sigma + \neg\psi)$$

$$\nvdash \neg Pr(\ulcorner \neg\psi \to \bar{0} = \bar{1}\urcorner), \quad \text{by definition of Con}$$

$$\nvdash \neg Pr(\ulcorner \psi \urcorner), \quad \text{by D1, D2,}$$

whence

$$\Sigma \nvdash \neg\psi \to \neg Pr(\ulcorner \psi \urcorner)$$

$$\nvdash Pr(\ulcorner \psi \urcorner) \to \psi.$$

Once this new proof came to be widely known, it was widely over-emphasized. Solovay (1976) named the modal system PrL "G" after Gödel, Boolos (1979) named his book on this modal logic based on Löb's Derivability Conditions (which he, like others, confused with Bernays') and Löb's Theorem after Gödel's Second Incompleteness Theorem, and moderate compromisers still call the logic GL for Gödel–Löb. No doubt this is done because of the glory attached to Gödel's name; it is not caused by *any* influence the Second Incompleteness Theorem had on the development of the subject and, as historians, we should be aware of this: Kripke's discovery of the proof was back in 1967, but by the time this proof became well known the new paradigms had already been discovered. This may be a bit of luck: Knowledge of this proof might have made Gödel's Second Incompleteness Theorem the paradigm and it might have taken longer for a Macintyre and Simmons to come along .... In this regard, note that Kreisel's preferred fixed point

$$\Sigma \vdash \varphi \leftrightarrow Pr(\ulcorner \varphi \to \psi \urcorner)$$

satisfies

$$\Sigma \vdash \varphi \leftrightarrow Pr(\ulcorner \neg \psi \to \neg \varphi \urcorner)$$

$$\vdash \neg \varphi \leftrightarrow \neg Pr_{\Sigma + \neg \psi}(\ulcorner \neg \varphi \urcorner),$$

i.e., $\neg \varphi$ is a Gödel sentence for $\Sigma + \neg \psi$, and we are left, á la Kreisel and Takeuti, with a too narrow self-referential paradigm. (One good thing: $\neg \varphi$ does not literally express its unprovability from $\neg \psi$).

But, if I complain of the over-invocation of Gödel's name in the modern work, this must be a minor complaint: After performing the sacred rite of reverence, perhaps crediting Löb's Theorem to Gödel or perhaps not, the modern researchers then go on their merry ways as if they had never heard of the Second Incompleteness Theorem or the paradoxes. They have looked to proof theory and universal algebra for non-self-referential paradigms for research in PrL. In short, a small, but healthy, development is in progress. We no longer see, as we did with Löb's sticking to a sentence φ expressing $Pr(\ulcorner \varphi \urcorner) \to \psi$ or with the textbook writers' failure to mention the Diagonalization Lemma, a turning of one's back on the obvious. Of course, this may be short-lived: As one wedded to the modal programme, which, as I have said but not convincingly indicated above, has begun to show some real progress, I find it hard to appreciate anything that does not fit my perspective—the usual complaint of maturity. For example, in *Gödel, Escher and Bach*, Douglas Hofstadter (1979) asked about *explicit* Henkin sentences, i.e., sentences which not only assert their own provability, but also provide what they consider to be proofs. He noted that, since

$$\Sigma \vdash \bar{0} = \bar{1} \leftrightarrow \text{Prov}(\bar{0}, \ulcorner \bar{0} = \bar{1} \urcorner),$$

$\bar{0} = \bar{1}$ is a refutable explicit Henkin sentence and asked if there were any provable explicit Henkin sentences, i.e., pairs φ, t such that

$$\Sigma \vdash \varphi \leftrightarrow \text{Prov}(t, \ulcorner \varphi \urcorner)$$

with $\Sigma \vdash \varphi$. Solovay (1985) answered this affirmatively. This is appealing, but I cannot say why and I thus ignored the result in my book (Smoryński, 1985). Do I get to say, along with one world leader, "History will absolve me," or will I qualify as an obscure indication that in 1985 the new concepts had not yet been universally recognized?

References

Van Bentham, J. (1978), Four paradoxes, *J. Philos. Logic* **7**, 49–72.

Bernardi, C. (1975), The fixed point theorem for diagonalizable algebras, *Studia Logica* **34**, 239–251.

Bernardi, C. (1976), The uniqueness of the fixed point in every diagonalizable algebra, *Studia Logica* **35**, 335–343.

Boolos, G. (1979), *The Unprovability of Consistency*, Cambridge University Press, Cambridge, UK.

Carlson, T. (1985), Modal logics with several operators and provability interpretations, *Israel J. Math.*, to appear.

Feferman, S. (1960), Arithmetization of metamathematics in a general setting, *Fund. Math.* **49**, 35–92.

Friedman, H. McAloon, K. and Simpson, S. (1982), A finite combinatorial principle which is equivalent to the 1-consistency of predicative analysis, in G. Metakides (ed.), *Patras Logic Symposion*, North-Holland, Amsterdam.

Gödel, K. (1931), Über formal unentscheidbare Sätze der *Principia Mathematica* und verwandter Systeme I, *Monatsh. Math. Phys.* **38**, 173–198.

Henkin L. (1952), Problem, *J. Symbolic Logic* **17**, 160.

Hilbert, D. and Bernays, P. (1939), *Grundlagen der Mathematik II*, Springer-Verlag, Berlin.

Hofstadter, D. (1979), *Gödel, Escher and Bach*, Harvester Press, Sussex, UK.

Kreisel, G. (1953), On a problem of Henkin's, *Proc. Netherlands Acad. Sci.* **56**, 405–406.

Kreisel, G. (1965), Mathematical logic, in T.L. Saaty (ed.), *Lectures on Modern Mathematics III*, Wiley, New York.

Kreisel, G. (1968), Notes concerning the elements of proof theory, Lecture Notes, UCLA.

Kreisel, G. Review of Macintyre and Simmons (1973), Zentralblatt 288.02018.

Kreisel, G. and Takeuti, G. (1974), Formally self-referential propositions for cut-free classical analysis and related systems, *Diss. Math.* **118**.

Löb, M.H. (1954), Solution of a problem by Leon Henkin, *Proc. of the ICM 1954 II*, 405–406, North-Holland, Amsterdam.

Löb, M.H. (1955), Solution of a problem of Leon Henkin, *J. Symbolic Logic* **20**, 115–118.

Macintyre, A. and Simmons, H. (1973), Gödel's diagonalization technique and related properties of theories, *Colloq. Math.* **28**, 165–180.

Montagna, F. (1978), On the algebraization of a Feferman's predicate, *Studia Logica* **37**, 221–236.

Montagna, F. (1984), The modal logic of provability, *Notre Dame J. Formal Logic* **25**, 179–189.

Montague, R. (1963), Syntactical treatments of modality, with corollaries on reflexion principles and finite axiomatizability, *Acta Phil. Fennica* **16**, 153–167.

Sambin, G. (1974), Un estensione del theorema di Löb, *Rend. Sem. Mat. Univ. Padova* **52**, 193–199.

Sambin, G. (1976), An effective fixed-point theorem in intuitionistic diagonalizable algebras, *Studia Logica* **35**, 345–361.

Simpson, S. (1983), Review of Paris, *J. Symbolic Logic* **48**, 482–483.

Smoryński, C. (1979), Calculating self-referential statements I: explicit calculations, *Studia Logica* **38**, 17–36.

Smoryński, C. (1981a), Review of Boolos (1979), *J. Symbolic Logic* **46**, 871–873.

Smoryński, C. (1981b), Fifty years of self-reference in arithmetic, *Notre Dame J. Formal Logic* **22**, 357–374.

Smoryński, C. (1985), *Self-Reference and Modal Logic*, Springer-Verlag, New York.

Smoryński, C. (1987), Quantified modal logic and self-reference, *Notre Dame J. Formal Logic* **28**, 356–370.

Solovay, R.M. (1976), Provability interpretations of modal logic, *Israel J. Math.* **25**, 287–304.

Solovay, R.M. (1985), Explicit Henkin sentences, *J. Symbolic Logic* **50**, 91–93.

The Unintended Interpretations of Intuitionistic Logic

Wim Ruitenburg

Abstract. We present an overview of the unintended interpretations of intuitionistic logic that arose after Heyting formalized the "observed regularities" in the use of formal parts of language, in particular, first-order logic and Heyting Arithmetic. We include unintended interpretations of some mild variations on "official" intuitionism, such as intuitionistic type theories with full comprehension and higher order logic without choice principles or not satisfying the right choice sequence properties. We conclude with remarks on the quest for a correct interpretation of intuitionistic logic.

1. The Origins of Intuitionistic Logic

Intuitionism was more than 20 years old before A. Heyting produced the first complete axiomatizations for intuitionistic propositional and predicate logic: according to L.E.J. Brouwer, the founder of intuitionism, logic is secondary to mathematics. Some of Brouwer's papers even suggest that formalization cannot be useful to intuitionism. One may wonder, then, whether intuitionistic logic should itself be regarded as an unintended interpretation of intuitionistic mathematics.

I will not discuss Brouwer's ideas in detail (on this, see Brouwer (1975) and Heyting (1934, 1956)), but some aspects of his philosophy need to be highlighted here. According to Brouwer mathematics is an activity of the human mind, a product of languageless thought. One cannot be certain that language is a prefect reflection of this mental activity. This makes language an uncertain medium (see van Stigt (1982) for more details on Brouwer's ideas about language).

In "De onbetrouwbaarheid der logische principes" ((Brouwer, 1981, pp. 253–259); for English translations of Brouwer's work on intuitionism, see Brouwer (1975)) Brouwer argues that logical principles should not guide but describe regularities that are observed in mathematical practice. The Principle of Excluded Third, $A \vee \neg A$, is an example of a logical principle that has become a guide for mathematical practice instead of simply describing it: the Principle of Excluded Third is observed in verifiable "finite" situations and generalized to a rule of mathematics. But according to Brouwer mathematics is not an experimental science, in which one only has to repeat an experiment sufficiently often to establish a law, so the Principle of Excluded Third should be discarded.

All his life Brouwer avoided the use of a formal language or logic, perhaps because of its unreliability, perhaps because of his personal style (see Brouwer (1981a, p. xi)). This does not imply that he did not believe in the possibility of a useful place for logic in intuitionistic mathematics, but rather that Brouwer would not himself resort to a formal language. This attitude was detrimental to the development of intuitionistic logic: it was not until 1923 that Brouwer discovered the equivalence in intuitionistic mathematics of triple negation and single negation (Brouwer, 1925).

While Brouwer may have been uncompromising with respect to his philosophy, his mathematical and philosophical talent was understood and appreciated by his thesis adviser D.J. Korteweg. In 1908 Korteweg advised Brouwer, after Brouwer completed his thesis, to devote some time to "proper" mathematics, as opposed to foundations, so as to earn recognition and become eligible for a university position. This initiated Brouwer's brief career as a topologist. Between 1909 and his appointment as Professor at the University of Amsterdam in 1912 Brouwer developed the technique of triangulation, showed the invariance of dimension, and proved his fixed-point theorem (Brouwer, 1975a, b).

Brouwer's ideas about language did not prevent others from considering formalizations of parts of intuitionism. A.N. Kolmogorov (1925) gave an incomplete description of first-order predicate logic. Of particular interest is his description of the double negation translation. Although this is sketchy in some parts, it is fair to say that Kolmogorov anticipated Gödel's translation from classical to intuitionistic formal systems (§4, Gödel, 1933). V. Glivenko (1928) described a fragment of intuitionistic propositional logic in order to establish the double negation of the Principle of Excluded Third. This was in reply to M. Barzin and A. Errera's 1927 paper, in which they attempted to prove Brouwer's mathematics inconsistent. Glivenko's 1929 paper appeared after he had seen Heyting's formalization of intuitionistic logic. In this paper Glivenko showed that the double negation of each classically derivable propositional statement is intuitionistically derivable. This result cannot be extended to first-order logic, as $\neg \forall x (P(x) \vee \neg P(x))$ cannot be contradicted in intuitionistic logic for unary atomic predicates P.

It took someone other than Brouwer to provide the first complete axiomatization of first-order intuitionistic logic. Heyting, a former student of Brouwer, wrote his Ph.D. thesis in 1925 on an intuitionistic axiomatization of projective geometry, the first substantial contribution to the intuitionistic program not from Brouwer himself (Troelstra, 1981). In 1927 the Dutch Mathematical Society published a prize question that included the quest for a formalization of intuitionistic mathematics. Heyting wrote an essay on the topic, for which he was awarded the prize in 1928. An expanded version appeared in Heyting (1930a, b, c). It included an axiomatization of intuitionistic first-order logic (using a version of equality between partial terms, thereby foreshadowing the existence predicate of D.S. Scott (1979)); Heyting Arithmetic, HA (the intuitionistic equivalent of Peano Arithmetic, PA); and analysis (the theory of choice sequences), though this last axiomatization was not complete.

In modern notation, using sequents, Heyting's axiomatization of intuitionistic predicate logic can be stated as follows. There are three logical axiom schemas and two axiom schemas for equality. The closure rules are: a thin horizontal line means that if the sequents above the line hold, then so do the ones below the line; a fat line means the same as a thin line, but in either direction.

$$A \vdash A$$

$$\frac{A \vdash B \quad B \vdash C}{A \vdash C}$$

$$A \vdash \top \qquad \qquad \bot \vdash A$$

$$\frac{A \vdash B \quad A \vdash C}{A \vdash B \wedge C} \qquad \frac{B \vdash A \quad C \vdash A}{B \vee C \vdash A}$$

$$\frac{A \wedge B \vdash C}{A \vdash B \to C}$$

$$\frac{A \vdash Bx}{A \vdash Bt} \dagger$$

$$\frac{A \vdash Bx}{A \vdash \forall x Bx} \ddagger \qquad \frac{Bx \vdash A}{\exists x Bx \vdash A} \ddagger$$

$$\top \vdash x = x$$

$$x = y \vdash Ax \to Ay \ *$$

In case $\dagger$ the variable x is not free in A and the term t does not contain a variable bound by a quantifier of B; in cases $\ddagger$ the variable x is not free in A; and in case $*$ the variable y is not bound by a quantifier of A.

Let Γ be a set of formulas, sequents, and rules and let σ be a formula. We write $\Gamma \vdash \sigma$, Γ *proves* σ, if there exists a finite subset $\Delta \subseteq \Gamma \cup \{\top\}$ such that $\bigwedge_{\delta \in \Delta} \delta \vdash \sigma$ is a sequent derivable from the system above plus the sequents and rules of Γ. We use $\vdash \sigma$ as an abbreviation for $\top \vdash \sigma$, $\neg \sigma$ as an abbreviation for $\sigma \to \bot$, and $\sigma \leftrightarrow \tau$ as an abbreviation for $(\sigma \to \tau) \wedge (\tau \to \sigma)$. If we add the axiom schema $\vdash A \vee \neg A$ to the system above of intuitionistic logic, we obtain an axiomatization for classical first-order logic.

In the language of HA we have the usual binary function symbols $+$ and $\cdot$, the unary function symbol S, and the constant symbol 0. The axiom system of HA has the following nonlogical axiom schemas and rule:

$$Sx = 0 \vdash \bot,$$

$$Sx = Sy \vdash x = y,$$

$$\vdash x + 0 = x,$$

$$\vdash x + Sy = S(x + y),$$

$$\vdash x \cdot 0 = 0,$$

$$\vdash x \cdot Sy = (x \cdot y) + x,$$

$$\frac{A(x) \vdash A(Sx)}{A(0) \vdash A(x)},$$

where x is not free in $A(0)$ in the rule above. The system HA is strong enough to prove essentially all number-theoretic results we find in any text on number theory. The main exceptions involve theorems related to incompleteness proofs of PA and statements that are true in PA only because of their logical form, like $\vdash \omega \vee \neg \omega$ for some undecidable statement ω. In fact, if $HA \vdash \sigma \vee \tau$, then $HA \vdash \sigma$ or $HA \vdash \tau$. Similarly, if $HA \vdash \exists x \sigma x$, then $HA \vdash \sigma m$ for some natural number m.

The systems of classical mathematics and intuitionistic mathematics diverge more noticeably when we consider the theory of real numbers or abstract mathematics.

In a letter to O. Becker in 1933, Heyting described how he came to his axiomatization essentially by going through the axioms and rules of the *Principia Mathematica* (Whitehead and Russell, 1925) and making a new system of axioms and rules out of the acceptable axioms. (The papers by Troelstra (1978, 1981) discuss in more detail the period around the formalization of intuitionistic logic.)

Like Brouwer, Heyting considered logic with mixed feelings. Heyting feared that his formalization would divert attention from the underlying issues (see Heyting (1930a, b, c), or the discussion of these papers by A.S. Troelstra (1978, 1981)). Later he even expressed disappointment in that his papers Heyting (1930a, b, c), had distracted the attention from the underlying ideas to the formal systems themselves. Those papers, then, do not represent a divergence between Heyting's ideas and Brouwer's. On the contrary, Brouwer clearly appreciated Heyting's contributions to clarify intuitionism (Troelstra, 1978).

Alongside intuitionism other schools of constructive mathematics developed. Most well known among these are finitism (Hilbert and Bernays, 1934), Markov (recursive) constructivism (Šanin, 1958; Markov, 1962), and Bishop's constructivism (Bishop, 1967). A discussion of these can be found in Troelstra (1977). Here it suffices to note that all these constructive schools settled on the same first-order logic, by means of different philosophies. Most of the essential differences involve second- or higher-order mathematical questions.

Differences also occur at the level of arithmetic; in particular, the Markov school allows the so-called Markov Principle to extend HA:

$$\forall n(An \vee \neg An) \wedge \neg\neg \exists nAn \vdash \exists nAn,$$

where A is a formula of arithmetic. The corresponding *rule* is weaker and already holds in HA:

$$\frac{HA \vdash \forall n(An \vee \neg An) \wedge \neg\neg \exists nAn}{HA \vdash \exists nAn}.$$

2. Interpretations for Propositional Logic

The main impact of Heyting's formalization of intuitionistic logic was its availability to a much wider audience of mathematicians and logicians. For the first time nonintuitionists could get a hold on intuitionism. This brought some vital intellectual blood into the development of intuitionism; before 1940 there were very few intuitionists, even in the Netherlands (Troelstra, 1981). Now nonintuitionists began to take steps to relate intuitionistic logic with other concepts, in particular, the theory of recursive functions and model theory. An early example of this is Gödel (1932) (see Gödel (1986) for all early work of Gödel). In this paper Gödel essentially constructed a countable properly descending sequence of logics T_i between intuitionistic propositional calculus and the classical propositional calculus such that each T_i is valid on just those linearly ordered Heyting algebras of length at most $i + 1$.

The first completeness theorem for the intuitionistic propositional calculus came from M.H. Stone (1937) and A. Tarski (1938) in 1937–38. They discovered that topological spaces form a complete set of models for the intuitionistic propositional calculus in the same way that Boolean spaces do for the classical propositional calculus: the completeness theorem for the classical propositional calculus based on Boolean spaces X works by assigning to each atom a clopen (closed and open) subset of X and by interpreting disjunction, conjunction, and negation as set-theoretic union, intersection, and relative complement. For intuitionistic propositional logic Stone and Tarski obtained a completeness theorem for topological spaces by assigning open sets to atoms and by interpreting disjunction, conjunction, and implication as in the classical case, except that when a resulting set was not open, it had to be replaced by its interior. See the remarks below on the relationship of this interpretation to an interpretation for $S4$.

The significance of Tarski and Stone's interpretation (geometrization of logic) is clear from our current perspective. It was too early for further development, however, as category theory did not appear until the 1940s, and sheaf theory had to wait nearly until the 1950s.

Gödel (1933b) embedded intuitionistic propositional logic into Lewis's modal propositional logic $S4$. The system $S4$ is a propositional logic with a modal operator, $\Box$. For the logical constants $\top$ and $\bot$ and the logical operators $\wedge$, $\vee$, $\rightarrow$, and $\neg$ we take the rules and axioms of classical propositional logic. The modal operator $\Box$, "provable," satisfies all substitution instances of the axioms

$$\Box A \vdash A,$$

$$\Box(A \rightarrow B) \vdash \Box A \rightarrow \Box B,$$

$$\Box A \vdash \Box\Box A,$$

and the rule

$$\frac{\vdash A}{\vdash \Box A}.$$

In Gödel (1933b) we actually find two closely related translations. One of Gödel's embeddings $A \mapsto A'$ is equivalent to the following inductively defined translation:

$$\top' = \Box\top;$$

$$\bot' = \Box\bot;$$

$$p' = \Box p, \quad p \text{ an atom;}$$

$$(A \wedge B)' = \Box(A' \wedge B');$$

$$(A \vee B)' = \Box(A' \vee B'); \quad \text{and}$$

$$(A \rightarrow B)' = \Box(A' \rightarrow B').$$

We can easily show that some of the $\Box$'s in the translation above are redundant; $S4 \vdash A' \leftrightarrow \Box(A')$. For this interpretation

$$\vdash \sigma \quad \text{if and only if} \quad S4 \vdash \sigma'.$$

Gödel proved only one direction: if σ holds intuitionistically, then its translation σ' holds in $S4$. The reverse implication was established by J.C.C. McKinsey and A. Tarski (1948).

A topological model for $S4$ consists of a set X provided with a topology. We assign a subset $[\![P]\!] \subseteq X$ to each atom P and extend the interpretation inductively to all propositions of the language by defining

$$[\![\top]\!] = X;$$

$$[\![\bot]\!] = \varnothing;$$

$$[\![A \wedge B]\!] = [\![A]\!] \cap [\![B]\!];$$

$$[\![A \vee B]\!] = [\![A]\!] \cup [\![B]\!];$$

$$[\![\neg A]\!] = [\![A]\!]^c, \quad \text{where } ^c \text{ stands for complement;}$$

$$[\![A \rightarrow B]\!] = [\![A]\!]^c \cup [\![B]\!]; \quad \text{and}$$

$$[\![\Box A]\!] = \text{Int}[\![A]\!], \quad \text{where Int stands for interior.}$$

Using Gödel's interpretation of intuitionistic logic into $S4$, we obtain the Stone–Tarski interpretation for intuitionistic logic: a topological model for intuitionistic propositional logic consists of a set X provided with a topology. We assign an open subset $[\![P]\!] \subseteq X$ to each atom P. The interpretation is extended inductively to all propositions of the language by defining

$$[\![\top]\!] = X;$$

$$[\![\bot]\!] = \varnothing;$$

$$[\![A \wedge B]\!] = [\![A]\!] \cap [\![B]\!];$$

$$[\![A \vee B]\!] = [\![A]\!] \cup [\![B]\!];$$

$$[\![\neg A]\!] = \mathrm{Int}([\![A]\!]^c); \quad \text{and}$$

$$[\![A \to B]\!] = \mathrm{Int}([\![A]\!]^c \cup [\![B]\!]).$$

We write $(X, [\![\cdot]\!]) \vDash \varphi$ if $[\![\varphi]\!] = X$. The completeness theorem is: for all sets of propositions $\Gamma \cup \{\varphi\}$,

$$\Gamma \vdash \varphi \Leftrightarrow \text{for all } (X, [\![\cdot]\!]), \text{ if } (X, [\![\cdot]\!]) \vDash \gamma \text{ for all } \gamma \in \Gamma, \text{ then } (X, [\![\cdot]\!]) \vDash \varphi.$$

Several mild variations on the equivalence between $S4$ and Gödel's translation have been discovered since 1933. There is, however, another modal logic of note, which is based on the provability operator in PA. This system, called G, was discovered by R.M. Solovay (1976) to be sufficient to characterize a certain form of provability in PA. The system G distinguishes itself from $S4$ by having all substitution instances of the axioms

$$\Box(A \to B) \vdash \Box A \to \Box B,$$

$$\Box(\Box A \to A) \vdash \Box A \qquad \text{(Löb's rule)},$$

and the rule

$$\frac{\vdash A}{\vdash \Box A}$$

for the modal operator $\Box$. The axiom schema $\Box A \vdash \Box \Box A$ is derivable in G. Note that $S4$ and G are relatively inconsistent.

The formulas of the modal language of G can be embedded into the language of number theory as follows. Let $Prov(\lceil \sigma \rceil)$ be the proof predicate of PA. Each map Φ that maps the atoms p of the modal language to sentences Φp of the language of PA can be extended to a map of all formulas by induction: $\Phi \top = \top$, $\Phi \bot = \bot$, $\Phi(A \circ B) = \Phi A \circ \Phi B$ for $\circ \in \{\wedge, \vee, \to\}$, $\Phi \neg A = \neg \Phi A$, and $\Phi(\Box A) = Prov(\lceil \Phi A \rceil)$. Then Solovay's result says

$$G \vdash \sigma \quad \text{if and only if} \quad PA \vdash \Phi \sigma \quad \text{for all } \Phi.$$

In 1978 R. Goldblatt (1978) showed how to interpret intuitionistic propositional logic in G. Combine the translation $A \mapsto A'$ with the translation $A \mapsto A^{\Diamond}$ from the modal language to itself, defined inductively by $p^{\Diamond} = p$, $\top^{\Diamond} = \top$,

$\perp^\Diamond = \perp, (A \circ B)^\Diamond = A^\Diamond \circ B^\Diamond$ for $\circ \in \{ \wedge, \vee, \rightarrow \}, (\neg A)^\Diamond = \neg A^\Diamond$, and $(\Box A)^\Diamond = A^\Diamond \wedge \Box (A^\Diamond)$. Goldblatt's result can then be stated as

$$\vdash \sigma \quad \text{if and only if} \quad G \vdash (\sigma')^\Diamond.$$

In his paper Goldblatt mentioned an earlier proof by A. Kuznetsov and A. Muzavitski.

Combining the results of Solovay and Goldblatt, we obtain

$$\vdash \sigma \quad \text{if and only if} \quad PA \vdash \Phi((\sigma')^\Diamond) \quad \text{for all } \Phi.$$

Most interpretations for propositional calculus were also extended to first-order logic interpretations (see §5).

3. Realizability

By the early 1930s there were two theories that dealt with the notion of constructive process: intuitionism and recursion theory.

In 1931 Heyting developed the proof-interpretation that accompanies his formalization of intuitionism. In Heyting (1934) we find both Heyting's interpretation and Kolmogorov's problem-interpretation of 1932, while in Hilbert and Bernays (1934) the concept of "incomplete communication" of a constructive statement is discussed. Brouwer himself was vague about the interpretation of the logical operators.

On the other hand, the notion of a recursive function was, by Church's thesis (1936), considered to be equivalent to that of an effectively computable function.

S.C. Kleene was the first to seriously consider the possibility of a more precise connection between the two, in particular, between HA and recursive function theory (Kleene, 1973). When an intuitionist claims $\exists n \sigma n$ there should be an effective process p to find n. When an intuitionist claims $\forall m \exists n \sigma(m, n)$ there must be a construction p such that for all numbers m, pm is an effective process for finding an n such that $\sigma(m, n)$. Kleene conjectured in 1940 that this p should embrace the existence of a recursive function f such that $\sigma(m, fm)$ holds for all m. This was not at all obvious at the time (Kleene, 1973).

In Hilbert and Bernays (1934) it was expounded that an intuitionistic statement like $\exists m \sigma m$ is an incomplete communication of a more involved statement giving m such that σm. In the same way a statement like $\sigma \vee \tau$ is an incomplete communication of a more involved statement either establishing σ or establishing τ. Kleene's idea was to somehow add this missing information to statements so that when he arrived at expressions like $\forall m \exists n \sigma(m, n)$ the added information would include the requested recursive function. Since formulas in HA are defined inductively by complexity, adding the missing information would have to be done by induction on the complexity of formulas. In early 1941 Kleene finally succeeded in bringing this idea to fruition, finding a version of it which encoded the missing information in natural numbers: number realizability (Kleene, 1945).

In number realizability missing information, encoded in numbers, is added in such a way that a formula σ is realizable by a number e, written $er\sigma$, if it is derivable in HA. The intended meaning of $er\sigma$ is that the number e encodes information "missing" from the communication that the formula σ is true intuitionistically. It is defined by induction on the complexity of formulas of the language of HA. This notion of realizability did not follow Heyting's proof-interpretation (Heyting, 1934, §7), which indicated a certain liberty in defining variations and extensions of realizability, a fact already employed in Kleene (1945).

According to the proof-interpretation an implication $\sigma \rightarrow \tau$ is established only if we provide a construction C that converts a proof of σ into a proof of τ. The Hilbert–Bernays version distinguishes itself by requiring that the construction C convert information establishing σ into information establishing τ. A key idea of number realizability was to use partial recursive functions instead of total recursive functions, which helped in extending the definition of realizability through implication. A total construction replies to all input. If the input is not information establishing σ, the reply may be meaningless. A partial construction need only provide information (establishing τ) from information establishing σ; on other data it may be undefined. It is not clear whether before 1941 anyone else was aware of the importance of this distinction.

Let $\langle x, y \rangle$ represent a recursive pairing funtion over HA, with recursive projection functions $p_1 x$ and $p_2 x$. Thus, $x = \langle p_1 x, p_2 x \rangle$, $p_1 \langle x, y \rangle = x$, and $p_2 \langle x, y \rangle = y$. Let $\{\cdot\} \cdot$ represent the Kleene bracket expression, where $\{x\}y$ is the result of applying the xth partial recursive function to the number y; write $!\{x\}y$ to indicate that $\{x\}y$ is defined. Since $\{x\}y$ is a partial function, it is more natural to describe it by a recursive relation. This is essentially Kleene's T-predicate. For our purposes we use the existence of a ternary relation $T(x, y, z)$ such that $T(x, y, z) \leftrightarrow \{x\}y = z$. All expressions mentioned above are expressible in the language of HA.

Number realizability is a translation $A \mapsto xrA$ for formulas not containing the variable x, which is defined inductively by:

$$xr\top = \top;$$

$$xr\bot = \bot;$$

$$xr(t = u) = t = u, \qquad t, u \text{ terms};$$

$$xr(A \wedge B) = p_1 xrA \wedge p_2 xrB;$$

$$xr(A \vee B) = (p_1 x = 0 \rightarrow p_2 xrA) \wedge (p_1 x \neq 0 \rightarrow p_2 xrB);$$

$$xr(A \rightarrow B) = \forall y(yrA \rightarrow !\{x\}y \wedge \{x\}yrB);$$

$$xr(\exists yAy) = p_2 xrA(p_1 x); \quad \text{and}$$

$$xr(\forall yAy) = \forall y(!\{x\}y \wedge \{x\}yrAy).$$

A formula is called *almost negative* if it is built up from formulas of the form $\exists x(t = u)$ using $\wedge$, $\rightarrow$, and $\forall$. The added information in the construction of the realizability translation embodies the assumption that all functions are recursive. This idea can be formalized by the axiom schema ECT_0, the *extended Church's thesis*:

$$\forall x(Ax \rightarrow \exists y B(x, y)) \vdash \exists z \forall x(Ax \rightarrow !\{z\}x \wedge B(x, \{z\}x)),$$

where A is an almost negative formula. The study of Kleene's number realizability culminated in the following results of Troelstra (1973, p. 196):

$$HA + ECT_0 \vdash A \leftrightarrow \exists x(x\mathbf{r}A)$$

and

$$HA + ECT_0 \vdash A \quad \text{if and only if} \quad HA \vdash \exists x(x\mathbf{r}A).$$

So if $HA \vdash \forall x \exists y \sigma(x, y)$, then there is a number e such that $HA \vdash e\mathbf{r}\forall x \exists y \sigma(x, y)$. Using the translation above, this implies

$$HA \vdash \forall x(!\{e\}x \wedge p_2(\{e\}x)\mathbf{r}\sigma(x, p_1(\{e\}x))).$$

So $f(x) = p_1(\{e\}x)$ is the recursive function sought after for HA extended with ECT_0:

$$HA + ECT_0 \vdash \forall x \sigma(x, f(x)).$$

Number realizability did explicate the Hilbert–Bernays interpretation of incomplete information, but it did not give a recursive witness for $\forall x \exists y \sigma(x, y)$-sentences over HA simpliciter. To make the method above work for HA alone, Kleene introduced a variation of number realizability consisting of a translation $A \mapsto x\mathbf{q}A$ for formulas not containing the variable x, defined inductively by:

$$x\mathbf{q}\top = \top;$$

$$x\mathbf{q}\bot = \bot;$$

$$x\mathbf{q}(t = u) = t = u, \quad t, u \text{ terms};$$

$$x\mathbf{q}(A \wedge B) = p_1 x\mathbf{q}A \wedge p_2 x\mathbf{q}B;$$

$$x\mathbf{q}(A \vee B) = (p_1 x = 0 \rightarrow A \wedge p_2 x\mathbf{q}A) \wedge (p_1 x \neq 0 \rightarrow B \wedge p_2 x\mathbf{q}B);$$

$$x\mathbf{q}(A \rightarrow B) = \forall y((A \wedge y\mathbf{q}A) \rightarrow !\{x\}y \wedge \{x\}y\mathbf{q}B);$$

$$x\mathbf{q}(\exists yAy) = A(p_1 x) \wedge p_2 x\mathbf{q}A(p_1 x); \quad \text{and}$$

$$x\mathbf{q}(\forall yAy) = \forall y(!\{x\}y \wedge \{x\}y\mathbf{q}Ay).$$

Using q-realizability, we can show

$$HA \vdash \forall x(Ax \rightarrow \exists y B(x, y))$$

implies

$$HA \vdash \forall x(Ax \rightarrow !\{e\}x \wedge B(x, \{e\}x)) \quad \text{for some } e,$$

where A is almost negative. Then $f(x) = \{e\}x$ is the function that Kleene conjectured in 1940.

Kleene's original conjecture went beyond HA to claim that if in any intuitionistic system that includes HA we prove $\vdash \forall x \exists y \sigma(x, y)$, then there should be a recursive function f such that $\vdash \forall x \sigma(x, fx)$. For that reason Kleene developed a formal system for Heyting's Analysis, the theory of choice sequences, and in 1957 extended realizability to it (see Kleene (1973) for further references).

For later work on and applications of realizability, see Troelstra (1973) and Beeson (1985).

The connection above between intuitionistic arithmetic and recursion theory suggests that there may be connections between computer science and intuitionism as well.

The computer science aspect of recursion theory focuses on complexity theory rather than on the class of all general recursive functions. Following a conjecture of S. Cook, S. Buss found subsystems of HA that allowed for an analog of Kleene's realizability, except that instead of representing all general recursive functions, the representable functions are the ones restricted to subclasses of the polynomial-complexity hierarchy.

Let Σ_0^p, Π_0^p, Σ_1^p, Π_1^p, Σ_2^p, Π_2^p, ... be the Meyer–Stockmeyer polynomial hierarchy of predicates (Stockmeyer, 1976), where $NP = \Sigma_1^p$, and co-$NP = \Pi_1^p$. Define $\square_i^p = PTC(\Sigma_{i-1}^p)$, the functions computed by a polynomial time Turing machine with an oracle for a Σ_{i-1}^p-predicate. The 0, 1-valued functions of $\square_i^p$ form the class Δ_i^p. Note that $P = \Delta_1^p$ is the class of polynomial complexity predicates, and $\square_1^p$ is the class of polynomial complexity functions.

Based on earlier work, Buss (1985a) considered subsystems $IS_2^1 \subseteq IS_2^2 \subseteq IS_2^3 \subseteq \cdots$ of HA, defined as follows: let $|x|$ be the length of the number x written in binary form, that is, $|x| = \lceil \log_2(x + 1) \rceil$, and let $\lfloor x/2 \rfloor$ be division of x by 2, rounded down. A quantifier is *bounded* if it occurs in the form $\forall x \leq t$ or $\exists x \leq t$ for some term t. A quantifier is *sharply bounded* if it occurs in the form $\forall x \leq |t|$ or $\exists x \leq |t|$. A formula is of bounded complexity Σ_i^b if it is equivalent to a formula in prenex form with i bounded-quantifier alternations with Σ on the outside, ignoring sharply bounded quantifiers. A formula is of hereditarily bounded complexity $H\Sigma_i^b$ if all its subformulas are also of complexity Σ_i^b. The systems IS_2^i are axiomatized by basic sets of axioms (depending on i) and the following induction schema for $H\Sigma_i^b$ formulas A:

$$A\left\lfloor \frac{x}{2} \right\rfloor \to Ax \vdash A(0) \to \forall x Ax.$$

Then the systems IS_2^i satisfy (Buss, 1985b):

$$IS_2^i \vdash \forall x_1 \ldots x_n \exists y A(x_1, \ldots, x_n, y)$$

implies

there exists a $\square_i^p$-function $f: \mathbf{N}^n \to \mathbf{N}$ such that

$A(m_1, \ldots, m_n, f(m_1, \ldots, m_n))$ is valid in $\mathbf{N}$, for all $m_1, \ldots, m_n$.

Conversely, for each $\Box_1^p$-function f there is a formula $A(\mathbf{x}, y)$ such that for all $\mathbf{m} \in \mathbf{N}^n$, $A(\mathbf{m}, f\mathbf{m})$ is valid in $\mathbf{N}$, and $IS_2^i \vdash \forall \mathbf{x} \exists y A(\mathbf{x}, y)$.

4. The Double Negation Translation, and the Dialectica Interpretation

Gödel initiated two more unintended interpretations, which for purposes of exposition I will treat in this section. In 1925 Kolmogorov gave an incomplete sketch of the so-called double-negation translation, which essentially consisted of doubly negating each subformula of a formula. This embedded the classical predicate calculus into the intuitionistic calculus. Gödel's translation of Gödel in 1933 is an example of a fully developed double negation translation for the case of arithmetic, including a precise translation for the quantifiers.

In modern notation the double negation translation is as follows:

$$\top^{\bullet} = \top;$$
$$\bot^{\bullet} = \bot;$$
$$P^{\bullet} = \neg\neg P, \quad P \text{ an atom};$$
$$(A \wedge B)^{\bullet} = A^{\bullet} \wedge B^{\bullet};$$
$$(A \vee B)^{\bullet} = \neg(\neg A^{\bullet} \wedge \neg B^{\bullet});$$
$$(A \rightarrow B)^{\bullet} = A^{\bullet} \rightarrow B^{\bullet};$$
$$(\forall x A)^{\bullet} = \forall x A^{\bullet}; \quad \text{and}$$
$$(\exists x A)^{\bullet} = \neg \forall x \neg A^{\bullet}.$$

Note that $\vdash \sigma^{\bullet} \leftrightarrow \neg\neg\sigma^{\bullet}$. Let C be the axiom schema of Excluded Third, i.e., $\vdash A \vee \neg A$. Then

$$C \vdash \sigma \quad \text{if and only if} \quad \vdash \sigma^{\bullet}.$$

Since in HA an atomic formula is equivalent to its double negation, this implies that each formula in HA, built up from the atoms using only negation, conjunction, and universal quantification, is equivalent to its double negation. So from Gödel's translation we obtain the result that for formulas σ of arithmetic, built up from the atoms using only negation, conjunction, and universal quantification,

$$PA \vdash \sigma \quad \text{if and only if} \quad HA \vdash \sigma.$$

Since in classical logic each first-order formula is equivalent to a formula using only negation, conjunction, and universal quantification, this implies that PA and HA are equiconsistent and that PA is embedded in HA.

The result above was discovered independently by Gentzen and Bernays in 1933.

In 1958 Gödel gave the Dialectica interpretation, in which logical complexity was replaced by the use of higher types. The language $L(\mathbf{N} - \mathbf{HA}^\omega)$ that it was originally designed for is a typed language with equality. The types are defined inductively: there is a bottom type N and for each pair of types r, s an operator type $(r)s$ which is a set of operators from r to s. The axiom of extensionality $\forall fg \in (r)s((\forall x \in r\, fx = gx) \to f = g)$ is not assumed from the outset: In the model HRO mentioned below the elements of $(r)s$ are essentially algorithms describing functions from r to s rather than functions themselves. For each type s there are variables x^s, y^s, ... New terms can be constructed from old ones by composition, provided that the types match.

The theory $\mathbf{N} - \mathbf{HA}^\omega$ is based on many-sorted intuitionistic predicate logic plus some defining axioms for particular constants: constants $0 \in N$ and $S \in (N)N$, and for all r, s, and t, constants $\Pi_{s,t} \in (s)(t)s$, $\Sigma_{r,s,t} \in ((r)(s)t)((r)s)(r)t$, and $R_s \in (s)((N)(s)s)(N)s$. Subscripts and superscripts will be suppressed when the meaning of the terms is unambiguous. We write xyz as abbreviation for $(xy)z$. For 0 and S we have the usual number-theoretic axioms of zero and successor. For Π, Σ, and R, $\Pi xy = x$, $\Sigma xyz = xz(yz)$, $Rxy0 = x$, and $Rxy(Sz) = y(Rxyz)z$. The constants Π and Σ are analogous to the terms $K = \lambda xy . x$ and $S = \lambda xyz . xz(yz)$ of lambda calculus (Barendregt, 1984, p. 31), and the elements k and s of a combinatory algebra (Barendregt, 1984, p. 90). The constants R are used to construct the primitive recursive functions ((Troelstra, 1973, pp. 51ff), (Barendregt, 1984, pp. 568ff)). By using the higher types we actually obtain a larger class of recursive functions.

There are several models for this theory mentioned in Troelstra (1973, Ch. 2) and an overview in Troelstra (1977, pp. 1026ff). One non-trivial example is the model HRO, the *hereditarily recursive operations*. Assign a domain $V_r \subseteq \mathbf{N}$ to each type r such that $V_N = \mathbf{N}$ and $x \in V_{(r)s} \leftrightarrow \forall y \in V_r \exists z \in V_s(\{x\}y = z)$. To make all sets V_r disjoint, replace $n \in V_r$ by $(n, r) \in V_r$. HRO is not extensional, as different elements (m, r), $(n, r) \in V_r$ may define the same partial recursive function $\{m\} = \{n\}$. Application is defined by $(m, (r)s) \cdot (n, r) = (\{m\}n, s)$.

The fundamental idea behind the Dialectica interpretation is the exchange of complexity for higher types. This is illustrated by the replacement of formulas of the form $\forall x \in r \exists y \in s\sigma(x, y)$ by $\exists z \in (r)s \forall x\sigma(x, zx)$. In the presence of sufficiently strong choice principles and Excluded Middle all formulas are equivalent to formulas in prenex form, and repeated application of the replacement above translates a formula A into an equivalent formula of the form $A^D = \exists \mathbf{x} \forall \mathbf{y} A_D(\mathbf{x}, \mathbf{y})$ with A_D quantifier-free. The Dialectica interpretation generalizes this translation from classical logic to an intuitionistic theory similar to $\mathbf{N} - \mathbf{HA}^\omega$, called $\mathbf{WE} - \mathbf{HA}^\omega$. The theory $\mathbf{WE} - \mathbf{HA}^\omega$ distinguishes itself from $\mathbf{N} - \mathbf{HA}^\omega$ by the schemas

$$x^{(r)s} = y^{(r)s} \leftrightarrow \forall z^r(x^{(r)s}z^r = y^{(r)s}z^r)$$

and

$$\frac{P \vdash \mathbf{xz} = y\mathbf{z} \vdash A\mathbf{x}}{\vdash Ay},$$

where P is a quantifier-free formula with terms of type N only in which x, y, and z do not occur. The above schemas make $\mathbf{WE} - \mathbf{HA}^\omega$ *weakly extensional*. For each formula Az we define two translations, A^D and A_D, where $A^D = \exists \mathbf{x} \forall \mathbf{y} A_D(\mathbf{x}, \mathbf{y}, \mathbf{z})$ and A_D is quantifier free. We extend the translation inductively as follows (n is a variable over N):

$$A^D = A_D = A, \qquad A \text{ an atom.}$$

For the other clauses, let $A^D = \exists \mathbf{x} \forall \mathbf{y} A_D(\mathbf{x}, \mathbf{y})$, $B^D = \exists \mathbf{u} \forall \mathbf{v} B_D(\mathbf{u}, \mathbf{v})$. Then:

$$(A \wedge B)_D = A_D \wedge B_D;$$

$$(A \vee B)_D = (n = 0 \rightarrow A_D) \wedge (n \neq 0 \rightarrow B_D);$$

$$(A \rightarrow B)_D = A_D(\mathbf{x}, \mathbf{Yxv}) \rightarrow B_D(\mathbf{Ux}, \mathbf{v});$$

$$(\exists z A(z))_D = A(z)_D;$$

$$(\forall z A(z))_D = A_D(\mathbf{X}z, \mathbf{y}, z);$$

$$(A \wedge B)^D = \exists \mathbf{xu} \forall \mathbf{yv} (A \wedge B)_D$$
$$= \exists \mathbf{xu} \forall \mathbf{yv} (A_D \wedge B_D);$$

$$(A \vee B)^D = \exists n \mathbf{xu} \forall \mathbf{yv} (A \vee B)_D$$
$$= \exists n \mathbf{xu} \forall \mathbf{yv} (n = 0 \rightarrow A_D(\mathbf{x}, \mathbf{y})) \wedge (n \neq 0 \rightarrow B_D(\mathbf{u}, \mathbf{v}));$$

$$(A \rightarrow B)^D = (\exists \mathbf{x} \forall \mathbf{y} A_D \rightarrow \exists \mathbf{u} \forall \mathbf{v} B_D)^D$$
$$= (\forall \mathbf{x} \exists \mathbf{u} \forall \mathbf{v} \exists \mathbf{y} (A_D \rightarrow B_D))^D$$
$$= \exists \mathbf{UY} \forall \mathbf{xv} (A \rightarrow B)_D$$
$$= \exists \mathbf{UY} \forall \mathbf{xv} (A_D(\mathbf{x}, \mathbf{Yxv}) \rightarrow B_D(\mathbf{Ux}, \mathbf{v}));$$

$$(\exists z A(z))^D = \exists z \mathbf{x} \forall \mathbf{y} (\exists z A(z))_D$$
$$= \exists z \mathbf{x} \forall \mathbf{y} A_D(\mathbf{x}, \mathbf{y}, z); \quad \text{and}$$

$$(\forall z A(z))^D = \exists \mathbf{X} \forall z \mathbf{y} (\forall z A(z))_D$$
$$= \exists \mathbf{X} \forall z \mathbf{y} A_D(\mathbf{X}z, \mathbf{y}, z).$$

A^D and A need not be intuitionistically equivalent unless A is of the form $\exists \mathbf{x} \forall \mathbf{y} B$ with B quantifier-free. The Dialectica interpretation has the following properties.

If $\mathbf{WE} - \mathbf{HA}^\omega \vdash A$ then $\mathbf{WE} - \mathbf{HA}^\omega \vdash \forall \mathbf{y} A_D(\mathbf{t}, \mathbf{y})$ for some sequence of terms $\mathbf{t}$,

and

$$\mathbf{WE} - \mathbf{HA}^\omega + S \vdash A \quad \text{if and only if} \quad \mathbf{WE} - \mathbf{HA}^\omega \vdash A^D$$

for particular extensions S of $\mathbf{WE} - \mathbf{HA}^\omega$ (Troelstra, 1977, pp. 1032ff). See Troelstra (1973, Ch. 3, §5) and (1977) for more on the Dialectica interpretation.

5. Interpretations for Predicate Logic

The extension of the Stone–Tarski models for propositional logic to first-order predicate logic was done by A. Mostowski (1948). More models for predicate logic soon followed: pseudo-Boolean algebras by McKinsey and Tarski (1948) and H. Rasiowa and R. Sikorski (1953); Beth models by E.W. Beth (1956, 1959); and Kripke models by S. Kripke (1965). Kripke models turned out to be particularly efficient and easy to use (see, for instance, (Smoryński, 1973)), because they implicitly use partial elements in their definition. I will give a description of Kripke models below, where I discuss a first-order extension of Lewis's system S4, leaving the others as special cases—up to isomorphism—of the sheaf models of §6. There were some techniques to convert a model of one kind to a model of another, but there was no unifying concept of model. In fact, classical set theory seems not to permit the construction of a *natural* unifying concept. The development of a unifying concept had to wait until the further advancement of category theory, particularly topos theory.

Gödel's interpretation of the intuitionistic propositional calculus in Lewis's system S4 was extended to an interpretation of the intuitionistic predicate calculus in QS4, the first-order generalization of S4, independently by Rasiowa and Sikorski (1953b) and S. Maehara (1954). Kripke models were announced in Kripke (1959) in 1959 as a specialization of Kripke's models for the system of modal predicate logic QS4, in which intuitionistic logic could be embedded.

The system QS4 is the first-order extension of the system S4, where for terms, equality, and the quantifiers $\exists$ and $\forall$, we have the usual axioms and closure rules:

$$\frac{A \vdash Bx}{A \vdash Bt} \ \dagger$$

$$\frac{A \vdash Bx}{A \vdash \forall x Bx} \ \ddagger \qquad \frac{Bx \vdash A}{\exists x Bx \vdash A} \ \ddagger$$

$$\top \vdash x = x$$

$$x = y \vdash Ax \to Ay \ *$$

In case $\dagger$ the variable x is not free in A and the term t does not contain a variable bound by a quantifier of B; in cases $\ddagger$ the variable x is not free in A; and in case $*$ the variable y is not bound by a quantifier of A.

Gödel's embedding $A \mapsto A'$ can be extended to first-order logic by

$$(\forall x Ax)' = \square(\forall x A'x); \quad \text{and}$$

$$(\exists x Ax)' = \square(\exists x A'x).$$

Again we have $QS4 \vdash A' \leftrightarrow \square(A')$, and

$$\vdash \sigma \quad \text{if and only if} \quad QS4 \vdash \sigma'.$$

Kripke models for $QS4$ are defined by pairs $\mathbf{K} = \langle S, I \rangle$, where $S = (\mathbf{P}, D_S)$ is an inhabited *structure* and I is an *interpretation*, as follows.

The notion of structure generalizes the traditional notion of set. A structure $S = (\mathbf{P}, D_S)$ consists of a partially ordered set $\mathbf{P} = (P, \leq)$—P is called the set of nodes—and a functor D_S from $\mathbf{P}$ to the category of sets. For an inhabited structure, the sets must be nonempty. The structure S must be inhabited for the same reason that models of first-order logic must be inhabited: it allows for simpler rules for the quantifiers. So for each $\alpha \in P$ we have a nonempty set $D_S\alpha$; for $\alpha \leq \beta$ a function $(\sigma_S)^\alpha_\beta: D_S\alpha \to D_S\beta$; and for $\alpha \leq \beta \leq \gamma$, $(\sigma_S)^\alpha_\gamma = (\sigma_S)^\beta_\gamma(\sigma_S)^\alpha_\beta$. In this way an element does not have to exist above all nodes. For each number $n \geq 0$, let $S^n = (\mathbf{P}, D_{S^n})$ be the structure on $\mathbf{P}$ defined by $(D_{S^n})\alpha = (D_S\alpha)^n$ and by $(\sigma_{S^n})^\alpha_\beta = ((\sigma_S)^\alpha_\beta)^n$. For all α $(D_{S^0})\alpha$ is a singleton. A substructure of S is a structure $R = (\mathbf{P}, D_R)$ on $\mathbf{P}$ such that $D_R\alpha \subseteq D_S\alpha$ for all nodes α and such that its maps $(\sigma_R)^\alpha_\beta$ are restrictions of the maps $(\sigma_S)^\alpha_\beta$. A map $F: S \to T$ between structures $S = (\mathbf{P}, D_S)$ and $T = (\mathbf{P}, D_T)$ consists of a collection of functions $\{F_\alpha: D_S\alpha \to D_T\alpha | \alpha \in \mathbf{P}\}$, such that for all $\alpha \leq \beta$ the following diagram commutes:

$$
\begin{array}{ccc}
D_S\alpha & \xrightarrow{\ F_\alpha\ } & D_T\alpha \\
{\scriptstyle (\sigma_S)^\alpha_\beta}\downarrow & & \downarrow{\scriptstyle (\sigma_T)^\alpha_\beta} \\
D_S\beta & \xrightarrow{\ F_\beta\ } & D_T\beta
\end{array}
$$

The structures and maps above form the functor category $\mathbf{S}^\mathbf{P}$, where $\mathbf{S}$ is the category of sets.

The interpretation I in the definition of Kripke model assigns to each n-ary atomic predicate P of the language a substructure R of S^n and to each n-ary function symbol f a map $F: S^n \to S$, where constant symbols are interpreted as 0-ary functions. The equality predicate is interpreted by the equality relation $E = (\mathbf{P}, D_E)$ on S^2 defined by $D_E\alpha = \{(e, e)|e \in S\alpha\}$. A term t is interpreted as the composition T of the interpretations of its parts.

Given a Kripke model $\mathbf{K} = \langle (\mathbf{P}, D), I \rangle$, let $L_\mathbf{K}$ be the extension of the first-order language L obtained by including constant symbols for the elements of $\coprod_{\alpha \in P} D\alpha$. Write $\alpha \Vdash_{QS4} P(t, u, \ldots)$, in words, $P(t, u, \ldots)$ is *satisfied above* α, if $(T_\alpha, U_\alpha, \ldots) \in R\alpha$. Extend the satisfaction relation $\Vdash_{QS4}$ inductively to all sentences of $L_\mathbf{K}$ as follows: expression $\alpha \Vdash_{QS4} \varphi$ is *well-formed* if all new constant symbols of $L_\mathbf{K}$ occurring in φ are from $D\alpha$. In that case we sometimes write φ_α instead of φ. Given $\alpha \leq \beta$ and φ_α, φ_β is the formula constructed from φ_α by replacing the constant symbols for $c \in D\alpha$ by the constant symbols for $\sigma^\alpha_\beta(c) \in D\beta$. Use the same notation for constant symbols and constants, as it is clear from context which interpretation is intended. Extend the definition of $\Vdash_{QS4}$ inductively by:

$$\alpha \Vdash_{QS4} \top;$$

$$\alpha\Vdash_{QS4}(\varphi \wedge \psi)_\alpha \Leftrightarrow \alpha\Vdash_{QS4}\varphi_\alpha \text{ and } \alpha\Vdash_{QS4}\psi_\alpha;$$

$$\alpha\Vdash_{QS4}(\varphi \vee \psi)_\alpha \Leftrightarrow \alpha\Vdash_{QS4}\varphi_\alpha \text{ or } \alpha\Vdash_{QS4}\psi_\alpha;$$

$$\alpha\Vdash_{QS4}(\varphi \rightarrow \psi)_\alpha \Leftrightarrow \alpha\Vdash_{QS4}\varphi_\alpha \text{ implies } \alpha\Vdash_{QS4}\psi_\alpha;$$

$$\alpha\Vdash_{QS4}(\vdash\varphi)_\alpha \Leftrightarrow \text{it is not the case that } \alpha\Vdash_{QS4}\varphi_\alpha;$$

$$\alpha\Vdash_{QS4}(\forall x\varphi(x))_\alpha \Leftrightarrow \alpha\Vdash_{QS4}\varphi(c)_\alpha \text{ for all } c \in D\alpha;$$

$$\alpha\Vdash_{QS4}(\exists x\varphi(x))_\alpha \Leftrightarrow \alpha\Vdash_{QS4}\varphi(c)_\alpha \text{ for some } c \in D\alpha; \quad \text{and}$$

$$\alpha\Vdash_{QS4}(\square\varphi)_\alpha \Leftrightarrow \beta\Vdash_{QS4}\varphi_\beta \text{ for all } \beta \geq \alpha.$$

We write $\mathbf{K} \vDash_{QS4} \varphi$ if $\alpha\Vdash_{QS4}\varphi_\alpha$ for all nodes $\alpha \in P$. The completeness theorem for $QS4$ then reads: for all sets of sentences $\Gamma \cup \{\varphi\}$,

$$QS4, \Gamma \vdash \varphi \Leftrightarrow \text{for all } \mathbf{K}, \text{ if } \mathbf{K} \vDash_{QS4} \gamma \text{ for all } \gamma \in \Gamma, \text{ then } \mathbf{K} \vDash_{QS4} \varphi.$$

Using the translation from intuitionistic predicate logic to $QS4$, we get the following interpretation for intuitionistic logic: Kripke models are pairs $\mathbf{K} = \langle S, I \rangle$ as before, with S an inhabited structure, and I assigning substructures, maps, and the equality relation to atomic predicates, function symbols, and the equality predicate. Define $\Vdash$ on atoms in exactly the same way as $\Vdash_{QS4}$. The extension of $\Vdash$ to the extended language of first-order intuitionistic predicate logic $L_\mathbf{K}$ differs, however: extend $\Vdash$ inductively by:

$$\alpha\Vdash\top;$$

$$\alpha\Vdash(\varphi \wedge \psi)_\alpha \Leftrightarrow \alpha\Vdash\varphi_\alpha \text{ and } \alpha\Vdash\psi_\alpha;$$

$$\alpha\Vdash(\varphi \vee \psi)_\alpha \Leftrightarrow \alpha\Vdash\varphi_\alpha \text{ or } \alpha\Vdash\psi_\alpha;$$

$$\alpha\Vdash(\varphi \rightarrow \psi)_\alpha \Leftrightarrow \beta\Vdash\varphi_\beta \text{ implies } \beta\Vdash\psi_\beta, \text{ for all } \beta \geq \alpha;$$

$$\alpha\Vdash(\forall x\varphi(x))_\alpha \Leftrightarrow \beta\Vdash\varphi(c)_\beta \text{ for all } \beta \geq \alpha \text{ and all } c \in D\beta; \quad \text{and}$$

$$\alpha\Vdash(\exists x\varphi(x))_\alpha \Leftrightarrow \alpha\Vdash\varphi(c)_\alpha \text{ for some } c \in D\alpha.$$

For predication, equality, negation, and bi-implication this means:

$$\alpha\Vdash P(t, u, \dots)_\alpha \Leftrightarrow (T_\alpha, U_\alpha, \dots) \in R_\alpha;$$

$$\alpha\Vdash(t = u)_\alpha \Leftrightarrow T_\alpha = U_\alpha;$$

$$\alpha\Vdash(\neg\varphi)_\alpha \Leftrightarrow \text{it is not the case that } \beta\Vdash\varphi_\beta \text{ for any } \beta \geq \alpha; \quad \text{and}$$

$$\alpha\Vdash(\varphi \leftrightarrow \psi)_\alpha \Leftrightarrow \text{for all } \beta \geq \alpha, \beta\Vdash\varphi_\beta \text{ if and only if } \beta\Vdash\psi_\beta.$$

We can easily verify that if $\alpha \leq \beta$ and $\alpha\Vdash\varphi_\alpha$, then $\beta\Vdash\varphi_\beta$.

Define $\mathbf{K} \vDash \varphi$ to mean that $\alpha\Vdash\varphi$ for all nodes $\alpha \in P$. The corresponding completeness theorem is: for all sets of sentences $\Gamma \cup \{\varphi\}$,

$$\Gamma \vdash \varphi \Leftrightarrow \text{for all } \mathbf{K}, \text{ if } \mathbf{K} \vDash \gamma \text{ for all } \gamma \in \Gamma, \text{ then } \mathbf{K} \vDash \varphi.$$

A major example of the use of Kripke models occurs in forcing as developed by P.J. Cohen in 1963 (Cohen, 1966). A generalized version is in Keisler (1973) and is based on work in Rasiowa and Sikorski (1963) on Boolean-valued model theory. The generalization includes the forcing methods introduced by A. Robinson and J. Barwise. One stage consists of constructing a so-called forcing property: given a countable language L with countably many constant symbols, construct a partially ordered set $\mathbf{P}$ where the nodes are pairs $\mathbf{p} = (p, f_p)$ such that the f_p are finite sets of atomic sentences from L with $f_p \subseteq f_q$ whenever $(p, f_p) \le (q, f_q)$. Construct a Kripke model $\mathbf{K}$ on $\mathbf{P}$ by assigning sets $D\mathbf{p} = \{c \mid \varphi c \in f_p$ for some $\varphi\}$ to $\mathbf{p} \in \mathbf{P}$, and relations R for each predicate n-ary P such that $R\mathbf{p} = \{\mathbf{c} \mid P\mathbf{c} \in f_p\}$, where $\mathbf{c} = (c_1, \ldots, c_n)$. The equality symbol is interpreted as a binary relation $\approx$ and need not be the standard equality relation of $\mathbf{K}$. Then $\mathbf{K}$ constitutes a *forcing property* if it satisfies

$$\mathbf{K} \models \forall x (\neg\neg x \approx x)$$

and

$$\mathbf{K} \models \forall xy((x \approx y \wedge \varphi(x)) \to \neg\neg\varphi(y))$$

for all formulas φ where y is not bound by a quantifier of φ.

Given a forcing property $\mathbf{K}$, the forcing relation $\Vdash$ is defined as for intuitionistic forcing, except that the equality symbol $=$ is interpreted by $\approx$. Weak forcing $\Vdash_\omega$ is defined by $\mathbf{p} \Vdash_\omega \varphi$ if and only if $\mathbf{p} \Vdash \varphi^*$, where φ^* is the double negation translation of φ. Forcing properties and the construction of generic models can now be considered as part of topos theory. The connection between topoi and independence proofs for the continuum hypothesis, was made by M. Tierney (1972). This was one of the main driving forces behind F.W. Lawvere and Tierney's development of topos theory.

6. Topoi

Topos theory has its origins in three separate lines of mathematical development. In the introduction of Johnstone (1977) we find the first and the third: sheaf theory and the category-theoretic foundation of mathematics. The second line is expounded in the preface of Goldblatt (1979).

The first line is sheaf theory, which was developed after the second world war as a tool for algebraic topology. Later the concept of a sheaf over a topological space was extended to that of a sheaf over a site in order to enable the construction of more "topologies" in algebraic geometry. Categories of sheaves over a site are known as Grothendieck topoi.

The second line of development was initiated by Cohen's forcing technique in his proof of the independence of the continuum hypothesis. It was realized very early that Cohen's forcing and Kripke's forcing were closely related examples of some common technique connecting intuitionistic logic and classical set theory. The topic was picked up by D.S. Scott and R. Solovay in 1965, when they developed the theory of Boolean-valued models of ZF. In the late 1960s Scott considered the natural generalization to Heyting-valued models.

The third line involved the attempts to axiomatize category-theoretically well-known categories such as module categories. P.T. Johnstone (Johnstone, 1977) mentioned as an early example the proof of the Lubkin–Heron–Freyd–Mitchell embedding theorem (Freyd, 1964) for abelian categories, which showed that there is an explicit set of elementary axioms that imply all the finitary exactness properties of module categories. Lawvere tried the same for the category of sets.

Cartesian closed categories preceded topoi. In fact, a topos is a finitely complete Cartesian closed category satisfying one extra property. A category is finitely complete if all finite diagrams have limits and colimits. A finitely complete Cartesian closed category $\mathbf{C}$ has all finite limits and colimits explicitly given by functors, and the functors $- \times b \colon \mathbf{C} \to \mathbf{C}$ have right adjoints. The category-theoretic axioms of forming new arrows from previously constructed ones are represented as sequent rules and axioms below. The letters $A, B, C, \dots$ represent object variables, and $f, g, h, \dots$ are arrow variables. The arrows 0_A, 1_A and id_A are introduced as axioms. A thin horizontal line means that if the arrows above the line exist, then there exists a corresponding arrow below the line. A fat line means the same as a thin line except that the correspondence is one-to-one and onto.

$$A \xrightarrow{\mathrm{id}_A} A$$

$$\frac{A \xrightarrow{f} B \quad B \xrightarrow{g} C}{A \xrightarrow{gf} C}$$

$$A \xrightarrow{1_A} 1 \qquad 0 \xrightarrow{0_A} A$$

$$\frac{A \xrightarrow{f} B \quad A \xrightarrow{g} C}{A \xRightarrow{\langle f,g \rangle} B \times C} \qquad \frac{B \xrightarrow{f} A \quad C \xrightarrow{g} A}{B \coprod C \xrightarrow{f+g} A}$$

$$\frac{A \times B \xrightarrow{f} C}{A \xrightarrow{\varphi f} C^B}$$

The notion of finitely complete Cartesian closed category is a straightforward generalization of intuitionistic propositional logic: take the formulas of a propositional logic as objects and have a unique arrow from φ to ψ exactly when $\varphi \vdash \psi$. Thus models of intuitionistic propositional logic give rise to examples of finitely complete Cartesian closed categories. This also implies that the notion of Cartesian closed category is too weak to capture a sufficient number of properties of higher order set theory.

Grothendieck topoi, on the other hand, do reflect a substantial part of set theory. A more detailed description of and references to the development of the notion of Grothendieck topos can be found in Gray (1979). Of main interest to us is Giraud's Theorem (1963–64), which characterizes Grothendieck topoi by certain exactness conditions (Johnstone, 1977, pp. 15–18). Lawvere considered Grothendieck topoi as models for his generalized set theory. His early axiomatization, however, still included some set-theoretic aspects, and was therefore not purely category-theoretical. An adaptation of Giraud's exactness conditions could provide a generalized set theory axiomatized purely in terms of finite exactness conditions and constructions. Lawvere's discovery that each Grothendieck topos has a subobject classifier $t: 1 \to \Omega$ completed that picture. It then turned out that a small number of finite exactness conditions, plus the existence of a subobject classifier, sufficed to develop a category-theoretic notion of set theory. During the year 1969–70 in Halifax Lawvere and Tierney developed the fundamentals of elementary topos theory.

An elementary *topos* is a finitely complete Cartesian closed category **E** that has a subobject classifier. The existence of finite colimits follows from the other axioms (Mikkelsen, 1976; Paré, 1974), so a topos **E** only has to satisfy:

(1) **E** has all finite products and equalizers.
(2) **E** is Cartesian closed: The functor $b \mapsto b \times a$ has a right adjoint functor $b \mapsto b^a$, for all a.
(3) **E** has a *subobject classifier* $t: 1 \to \Omega$. That is, for each monomorphism $f: a \to b$ there is a unique χ_f, the *classifying map*, from b to the *truth value object* Ω such that the following is a pullback:

In the definition above we assume that all limits are given functorially. For (1), finite products and equalizers are sufficient to construct all finite limits (Mac Lane, 1971, p. 109). For (2), the objects b^a behave like the function sets b^a in the category **S** of sets. The map χ_f of (3) then behaves like the characteristic function of the image of f. In fact, the category of sets **S** is a topos, with $\Omega \cong \{0, 1\}$.

Topoi have an internal logical structure just like **S**: intuitionistic type theory. This was first made explicit by W. Mitchell (1972). A topos need not satisfy any additional choice principles, like Dependent Choice or Countable Choice. It does however have full comprehension for each object, and a power-set construction: for each A, Ω^A is its power-object. Both category theory and intuitionism worked in a field that in some way extended the traditional universe of classical logic and sets. That both fields meet in topos theory suggests that they satisfy S. Mac Lane's dictum: good general theory does not search for the maximum generality, but for the right generality (Mac Lane, 1971, p. 103).

Colimits are derivable from the definition above for the same reason that existential quantification and disjunction (and, in fact, conjunction and negation) are definable in terms of universal quantification and implication over Ω (Prawitz, 1965; Scott, 1979):

$$\vdash (p \wedge q) \leftrightarrow \forall r((p \to (q \to r)) \to r),$$

$$\vdash (p \vee q) \leftrightarrow \forall r(((p \to r) \wedge (q \to r)) \to r),$$

$$\vdash (\neg p) \leftrightarrow \forall r(p \to r),$$

$$\vdash (\exists x \varphi(x)) \leftrightarrow \forall r(\forall x(\varphi(x) \to r) \to r),$$

where p, q, and r are variables over the set of truth values Ω. Similarly, the union of two subsets X and Y of a set S is the intersection of all the subsets of S containing both X and Y.

Let C be a small category. Then the functor category $\mathbf{S}^\mathbf{C}$ is a topos. Objects of $\mathbf{S}^\mathbf{C}$ are also called *presheaves*. So Kripke models are presheaves. Finite limits and colimits are created pointwise: $(X \times Y)(a) = X(a) \times Y(a)$ for all a; $1(a) = \{0\}$ for all a; and if F is the coequalizer of $f, g: X \to Y$, then for all a $F(a)$ is the coequalizer of $f(a)$ and $g(a)$. Let X and Y be presheaves of $\mathbf{S}^\mathbf{C}$. For the exponent presheaf Y^X we consider "local" natural transformations. Just taking function sets $Y^X(a) = Y(a)^{X(a)}$ usually does not work: functions at node a must be provided with information as to what they look like at later stages x, reached from a by maps $f: a \to x$. Let a be an object of C. The comma category $\mathbf{C} \uparrow a$ has as objects arrows $f: a \to x$ and $g: a \to y$, and as arrows maps $w: f \to g$, with $w: x \to y$ from C, satisfying $wf = g$. There are induced functors X_a and Y_a from $\mathbf{C} \uparrow a$ to S defined by $X_a(f) = X(x)$, $X_a(w) = X(w)$, $Y_a(f) = Y(x)$, and $Y_a(w) = Y(w)$. Set $Y^X(a)$ equal to the set of natural transformations from X_a to Y_a. For $p: b \to a$, define $Y^X(p): Y^X(b) \to Y^X(a)$ by $Y^X(\rho)_w = \rho_{wp}$. As to the truth value object Ω, let a be an object of C. Set $\Omega(a)$ to be the set of subobjects **s** in $\mathbf{S}^\mathbf{C}$ of the representable functor $\mathbf{C}(a, -): \mathbf{C} \to \mathbf{S}$. So $s(x) \subseteq \mathbf{C}(a, x)$ for all objects and arrows x of C. For $p: a \to b$, define $\Omega(p): \Omega(a) \to \Omega(b)$ by $\Omega(p)(\mathbf{s}) = \mathbf{t}$ with $f \in \mathbf{t}$ if and only if $fp \in \mathbf{s}$.

It is possible to construct subtopoi of *sheaves* $sh_j(\mathbf{E})$ of a topos **E** by restricting the number of possible truth values of Ω using a topology j, not to be confused with, and quite different from, a topology on a set. A Grothendieck topos is a category equivalent to a topos of the form $sh_j(\mathbf{S}^\mathbf{C})$, so Grothendieck topoi are elementary topoi. All models of first-order intuitionistic logic mentioned in §5 are sheaves in Grothendieck topoi.

The definition of topology went through several generalizations until Lawvere arrived at the following definition: a *topology* on a topos **E** is a map $j: \Omega \to \Omega$ satisfying (1) $j \cdot t = t$; (2) $j \cdot j = j$; and (3) $j \cdot \wedge = \wedge \cdot (j \times j)$, where $\wedge$ is the intersection map, the classifying map of $\langle t, t \rangle: 1 \to \Omega \times \Omega$. The subobject Ω_j of Ω of fixed points of j represents the remaining possible truth values and is the truth value object in $sh_j(\mathbf{E})$.

Rather than describe the general procedure of making a Grothendieck topos, we illustrate the construction for the special case of sheaves over a topological space. Let X be a topological space, and let $\mathbf{S^P}$ be the topos of presheaves on the partially ordered set $\mathbf{P} = O(X)^{op}$. The set $O(X)$ is the partially ordered set of open subsets of X ordered by inclusion. The dual category $\mathbf{P}$ is therefore the lattice of open subsets of X ordered by containment $U \leq V$ if and only if $U \supseteq V$. Presheaves S consist of sets $S(U)$ for all $U \in O(X)$ and restriction maps $\sigma_V^U: S(U) \to S(V)$ for all pairs of open sets $U \supseteq V$. The truth value object Ω is defined by $\Omega(U) = \{S \subseteq O(U) | W \subseteq V \in S$ implies $W \in S\}$. Define a topos-theoretic topology j that relates to the set-theoretic topology $O(X)$ by setting maps $j_U: \Omega(U) \to \Omega(U)$ such that $j_U(S) = \{V \in O(U)| V \subseteq \cup S\}$. Then Ω_j is such that $\Omega_j(U) = O(U)$ and $(\Omega_j)_V^U(W) = V \cap W$. The resulting topos $\mathrm{sh}_j(\mathbf{S^P}) = \mathrm{sh}(X)$ is the subcategory of sheaves of $\mathbf{S^P}$, that is, the presheaves R satisfying:

(1) If $S \subseteq O(U)$ is such that $\cup S = U$, and $x, y \in R(U)$ are such that $\sigma_V^U(x) = \sigma_V^U(y)$ for all $V \in S$, then $x = y$.
(2) If $S \subseteq O(U)$ is such that $\cup S = U$, and there are elements $x_V \in R(V)$ such that $\sigma_{V \cap W}^V(x_V) = \sigma_{V \cap W}^W(x_W)$ for all $V, W \in S$, then there exists $x \in R(U)$ such that $\sigma_V^U(x) = x_V$ for all $V \in S$.

Grothendieck topoi also have natural number objects. In functor categories $\mathbf{S^P}$ the constant presheaf $\mathbf{N}$ defined by $\mathbf{N}(a) = \omega$ for all a performs this role. Lawvere's original definition of natural number object is equivalent to P. Freyd's elementary characterization (Freyd, 1972): A topos has a *natural number object* $\mathbf{N}$ it there exist arrows $o: 1 \to \mathbf{N}$ and $s: \mathbf{N} \to \mathbf{N}$ (zero and successor) such that $1 \overset{o}{\to} \mathbf{N} \overset{s}{\leftarrow} \mathbf{N}$ is a coproduct diagram and $\mathbf{N} \to 1$ is a coequalizer of s and $\mathrm{id}_N: \mathbf{N} \to \mathbf{N}$. In Grothendieck topoi the natural number object satisfies all first-order statements of classical number theory, but in the general situation of elementary topoi with natural number object it only has to satisfy the higher order equivalent of HA.

Cohen's forcing and many methods used in independence proofs of set theory are in fact topos theoretic techniques (Tierney, 1972; Ščedrov, 1984). There are many more applications of internal intuitionistic logic via topos theory. Proceedings like Fourman et al. (1979) and Troelstra and van Dalen (1982), and monographs like Kock (1981) and Ščedrov (1984) show just a few examples of the possible applications of intuitionistic logic and topos theory to such areas as Banach spaces, analysis, sheaf theory, topology, differential geometry, complex variables, algebra, and set theory. The applications to classical mathematics confirm in a concrete way that proving something constructively really means proving something more.

Brouwer provided classical proofs for his results in topology and later denounced these proofs as being insufficient for an intuitionist. It will be ironic if intuitionistic methods and sheaf models provide a method to prove Brouwer's contributions to topology.

It was the discovery of the effective topos *Eff* by M. Hyland and the subsequent development of tripos theory that gave rise to topoi, which bring into topos theory the previous models of realizability and the Dialectica interpretation (Hyland, 1982; Hyland et al., 1980). In an unpublished manuscript in 1977, W. Powell formulated in a classical context a semantics for realizability that has analogies with Hyland's approach. A significant difference between old style realizability and *Eff*-style topoi is that realizability gives interpretations of logical structures, while *Eff* allows us to think of realizability as model theory. The truth value of a sentence A in *Eff* is the set of numbers e such that erA.

Currently topos theory is the unifying concept behind the unintended interpretations of intuitionistic logic.

7. Intended Interpretations of Intuitionistic Logic

With so many unintended interpretations available for intuitionistic logic, we might at first expect that there should be at least one undisputed proper interpretation. That this is not the case can be explained as follows: an unintended interpretation presents no reason for dispute since it is, after all, unintended. An intended interpretation has to reflect how a *real* intuitionist—Brouwer, say—interprets the connectives. Since Brouwer never made an attempt at this himself, we seem to get closest to it by considering Heyting's proof interpretation for first-order logic and its extensions by G. Kreisel.

The following describes the Brouwer–Heyting–Kreisel (BHK) proof interpretation. A statement φ is true only if we have a proof p for it which satisfies the following requirements:

(1) p proves $\varphi \wedge \psi$ just in case p consists of a pair q, r of proofs of φ and ψ, respectively.
(2) p proves $\varphi \vee \psi$ just in case p consists of a pair n, q such that either $n = 0$ and q proves φ or $n = 1$ and q proves ψ.
(3) p proves $\varphi \to \psi$ just in case p consists of a pair q, r such that q is a construction that converts each proof s of φ into a proof $q(s)$ of ψ and such that r is a proof that q is such a construction.
(4) p proves $\exists x \varphi(x)$ just in case p consists of a pair q, r such that q is a construction that yields an element c such that r is a proof of $\varphi(c)$.
(5) p proves $\forall x \varphi(x)$ just in case p consists of a pair q, r such that for all c in the domain, $q(c)$ is a proof of $\varphi(c)$, and such that r is a proof that q is so.

Kreisel (1962) proposed the addition of the "extra proof r" clauses in the descriptions of the $\to$-case and the $\forall$-case. The interpretation is not reductive: it does not break proofs p down into simpler notions. Parts of the proof interpretation have been brought into question (for references, see van Dalen (1982, p. 61)). Questions arose as to whether one can quantify over a universe of *all* proofs, or whether the extra proof r is of a nature similar to proofs p. Intuitionists see the proof interpretation as an *explanation* rather than an interpretation of intuitionistic logic. Moreover, as Brouwer indicated, (formal) language is not trustworthy, and from that point of view the dispute is not surprising. Heyting's formalization itself, however, appears to be undisputed.

Acknowledgments

This chapter would not have been completed had it not been for the availability of many excellent papers on historical aspects of constructive mathematics. In particular I wish to mention, and highly recommend, the papers of Troelstra on historical aspects of intuitionism, the historical appendix in Beeson (1985), and Kleene's (1973) autobiographical notes. I want to thank John Simms and Paul Bankston for many helpful comments and suggestions for improvement.

References

Barendregt, H.P. (1984), *The Lambda Calculus, Its Syntax and Semantics*, rev. ed., Studies in Logic and the Foundations of Mathematics, Vol. 103, North-Holland, Amsterdam.

Barwise, J. (ed.) (1977), *Handbook of Mathematical Logic*, Studies in Logic and the Foundations of Mathematics, Vol. 90, North-Holland, Amsterdam.

Barzin, M. and Errera, A. (1927), Sur la logique de M. Brouwer, *Acad. Roy. Belg. Bull. Cl. Sci.* **5**, No. 13, 56–71.

Beeson, M.J. (1985), *Foundations of Constructive Mathematics*, Ergebnisse der Mathematik und ihrer Grenzgebiete 3. Folge, Band 6, Springer-Verlag, Berlin.

Bertin, E.M.J., Bos, H.J.M. and Grootendorst, A.W. (eds.), (1978), *Two Decades of Mathematics in the Netherlands 1920–1940, Part I*. Centre for Mathematics and Computer Science, Amsterdam.

Beth, E.W. (1956), Semantic construction of intuitionistic logic, *Med. Konink. Nederl. Akad. Wetensch.*, new series **19**, No. 11.

Beth, E.W. (1959), *The Foundations of Mathematics*, North-Holland, Amsterdam.

Bishop, E. (1967), *Foundations of Constructive Analysis*, McGraw-Hill, New York.

Brouwer, L.E.J. (1925), Intuitionistische Zerlegung mathematischer Grundbegriffe, *Jahresber. Deutsch. Math.-Verein.* **33**, 251–256.

Brouwer, L.E.J. (1975a), *Collected Works*, Vol. I, *Philosophy and Foundations of Mathematics*, A. Heyting (ed.), North-Holland, Amsterdam.

Brouwer, L.E.J. (1975b), *Collected Works*, Vol. II, *Geometry, Analysis, Topology, and Mechanics*, H. Freudenthal (ed.), North-Holland, Amsterdam.

Brouwer, L.E.J. (1981a), Over de Grondslagen der Wiskunde, Ph.D. 1907, with additions from D.J. Korteweg, G. Mannoury, and D. van Dalen (ed.), (1981a), *CWI Varia*, Vol. 1, Centre for Mathematics and Computer Science, Amsterdam.

Brouwer, L.E.J. (1981b), *Brouwer's Cambridge Lecture Notes*, with preface and notes by D. van Dalen (ed.), Cambridge University Press, Cambridge, UK.

Buss, S.R. (1985a), Bounded Arithmetic, Ph.D. Thesis.

Buss, S.R. (1985b), The polynomial hierarchy and intuitionistic bounded arithmetic, preprint.

Cohen, P.J. (1966), *Set Theory and the Continuum Hypothesis*, Benjamin, New York.

Crossley, J.N. Dummett, M.A.E. (eds.), (1965), *Formal Systems and Recursive Functions*, North-Holland, Amsterdam.

van Dalen, D. (1982), Braucht die konstruktive Mathematik Grundlagen?, *Jahresber. Deutsch. Math.-Verein.* **84**, 57–78.

Fourman, M.P. Mulvey, C.J. Scott, D.S. (eds.), (1979), *Applications of Sheaves*, Lecture Notes in Mathematics 753, Springer, New York.

158   Wim Ruitenburg

Freyd, P. (1964), *Abelian Categories*, Harper and Row, New York.
Freyd, P. (1972), Aspects of topoi, *Bull. Austral. Math. Soc.* **7**, 1–76.
Glivenko, V. (1928), Sur la logique be M. Brouwer, *Acad. Roy. Belg. Bull. Cl. Sci.* **5**, No. 14, 225–228.
Glivenko, V. (1929), Sur quelques points de la logique de M. Brouwer, *Acad. Roy. Belg. Bull. Cl. Sci.* **5**, No. 15, 183–188.
Gödel, K. (1932), Zum intuitionistischen Aussagenkalkül, *Anzeiger der Akademie der Wissenschaften in Wien*, **69**, 65–66.
Gödel, K. (1933a), Zur intuitionistischen Arithmetik und Zahlentheorie, *Ergebnisse eines mathematischen Kolloquiums* **4**, 34–38.
Gödel, K. (1933b), Eine Interpretation des intuitionistischen Aussagenkalküls, *Ergebnisse eines mathematischen Kolloquiums* **4**, 39–40.
Gödel, K. (1958), Über eine bisher noch nicht benützte Erweiterung des finiten Standpunktes, *Dialectica* **12**, 280–287.
Gödel, K. (1986), *Kurt Gödel, Collected Work*, Vol. I, S. Feferman, J.W. Dawson, S.C. Kleene, G.H. Moore, R.M. Solovay, J. van Heijenoort (eds.), Oxford University Press/Clarendon Press, Oxford.
Goldblatt, R. (1978), Arithmetical necessity, provability, and intuitionistic logic, *Theoria* **44**, 38–46.
Goldblatt, R. (1979), *Topoi, the Categorial Analysis of Logic*, Studies in Logic and the Foundations of Mathematics, vol. 98, North-Holland, Amsterdam.
Gray, J.W. (1979), Fragments of the history of sheaf theory, in Fourman et al. (1979), pp. 1–79.
Heyting, A. (1930a), *Die formalen Regeln der intuitionistischen Logik*, Sitzungsberichte der preußischen Akademie von Wissenschaften, Physikalisch–mathematische Klasse, pp. 42–56.
Heyting, A. (1930b), *Die formalen Regeln der intuitionistischen Mathematik II*, Sitzungsberichte der preußischen Akademie von Wissenschaften, Physikalisch–mathematische Klasse, pp. 57–71.
Heyting, A. (1930c), *Die formalen Regeln der intuitionistischen Mathematik III*, Sitzungsberichte der preußischen Akademie von Wissenschaften, Physikalisch–mathematische Klasse, pp. 158–169.
Heyting, A. (1934), *Mathematische Grundlagenforschung. Intuitionismus. Beweistheorie*, Ergebnisse der Mathematik und ihrer Grenzgebiete 3. Folge, Band 4, Springer-Verlag, Berlin.
Heyting, A. (1956), *Intuitionism, an Introduction*, Studies in Logic and the Foundations of Mathematics, Vol. 34, North-Holland, Amsterdam.
Hilbert, D. and Bernays, P. (1934), *Grundlagen der Mathematik*, Vol. 1, Springer-Verlag, Berlin.
Hyland, J.M.E. (1982), The effective topos, in Troelstra and van Dalen (1982), pp. 165–216.
Hyland, J.M.E., Johnstone, P.T. and Pitts, A.M. (1980), Tripos theory, *Math. Proc. Camb. Philos. Soc.* **88**, 205–232.
Johnstone, P.T. (1977), *Topos Theory*, L.M.S. Monographs, Vol. 10, Academic Press, New York.
Keisler, H.J. (1973), Forcing and the omitting types theorem, in Morley (1973), pp. 96–133.
Kleene, S.C. (1945), On the interpretation of intuitionistic number theory, *J. Symbolic Logic* **10**, 109–124.

Kleene, S.C. (1973), Realizability: A retrospective survey, in Mathias and Rogers (1973), pp. 95–112.

Kock, A. (1981), *Synthetic Differential Geometry*, London Mathematical Society Lecture Note Series, Vol. 51, Cambridge University Press, Cambridge, UK.

Kolmogorov, A.N. (1925), O principe tertium non datur (Sur le principe de tertium non datur), *Mat. Sbornik* (Recueil mathematique de la societé mathematique de Moscou) 32, 646–667. (English translation in van Heijenoort (1967))

Kreisel, G. (1962), Foundations of intuitionistic logic, in Nagel et al. (1962), pp. 198–210.

Kripke, S.A. (1959), Semantical analysis of modal logic, *J. Symbolic Logic* 24, 323–324.

Kripke, S.A. (1965), Semantical analysis of intuitionistic logic I, in Crossley and Dummett (1965), pp. 92–130.

Lawvere, F.W. (ed.), (1972), *Toposes, Algebraic Geometry and Logic*, Lecture Notes in Mathematics 274, Springer-Verlag, New York.

Mac Lane, S. (1971), *Categories for the Working Mathematician*, Graduate Texts in Mathematics, vol. 5, Springer-Verlag, New York.

Maehara, S. (1954), Eine Darstellung der intuitionistischen Logik in der Klassischen, *Nagoya Math. J.* 7, 45–64.

Markov, A.A. (1962), On constructive mathematics, *Trudy Mat. Inst. Steklov* 67, 8–14; English translation in Amer. Math. Soc. Transl. 98, 2nd series (1971), 1–9.

Mathias, A.R.D. and Rogers, H. (eds.), (1973), *Cambridge Summer School in Mathematical Logic*, Lecture Notes in Mathematics 337, Springer-Verlag, New York.

McKinsey, J.C.C. and Tarski, A. (1948), Some theorems about the sentential calculi of Lewis and Heyting, *J. Symbolic Logic* 13, 1–15.

Mikkelsen, C.J. (1976), *Lattice-Theoretic and Logical Aspects of Elementary Topoi*, Århus Universitet Various Publications 25, Aarhus University.

Mitchell, W. (1972), Boolean topoi and the theory of sets, *J. Pure Appl. Algebra* 2, 261–274.

Morley, M.D. (ed.), (1973), *Studies in Model Theory*, MAA Studies in Mathematics 8, The Mathematical Association of America, New York.

Mostowski, A. (1948), Proof on non-deducibility in intuitionistic functional calculus, *J. Symbolic Logic* 13, 204–207.

Nagel, E., Suppes, P. and Tarski, A. (eds.), (1962), *Logic Methodology, and Philosophy of Science*, Stanford University Press, Stanford, CA.

Paré, R. (1974), Colimits in topoi, *Bull. Amer. Math. Soc.* 80, 556–561.

Prawitz, D. (1965), *Natural Deduction, A Proof-Theoretical Study*, Almqvist & Wiksell, Stockholm.

Rasiowa, H. and Sikorski, R. (1953a), On satisfiability and deducibility in non-classical functional calculi, *Bull. Acad. Polon. Sci. Cl.* 3, 229–231.

Rasiowa, H. and Sikorski, R. (1953b), Algebraic treatment of the notion of satisfiability, *Fund. Math.* 40, 62–95.

Rasiowa, H. and Sikorski, R. (1963), *The Mathematics of Metamathematics*, PWN—Polish Scientific Publishers, Warsaw.

Šanin, N.A. (1958), On the constructive interpretation of mathematical judgments, *Trudy Mat. Inst. Steklov* 52, 226–311; English translation in Amer. Math. Soc. Transl. 23, 2nd series (1963), 109–189.

Ščedrov, A. (1984), *Forcing and Classifying Topoi*, Mem. Amer. Math. Soc., Vol. 48, Nr. 295, American Mathematical Society, Providence, RI.

Scott, D.S. (1979), Identity and existence in intuitionistic logic, in Fourman et al. (1979), pp. 660–696.

Smoryński, C. (1973), Applications of Kripke models, in Troelstra (1973), pp. 324–391.

Solovay, R.M. (1976), Provability interpretations of modal logic, *Israel J. Math.* **25**, 287–304.

van Stigt, W.P. (1982), L.E.J. Brouwer, the signific interlude, in Troelstra and van Dalen (1982), pp. 505–512.

Stockmeyer, L.J. (1976), The polynomial-time hierarchy, *Theoret. Comput. Sci.* **3**, 1–22.

Stone, M.H. (1937), Topological representation of distributive lattices and Brouwerian logics, *Casopis Pest. Mat. Fys.* **67**, 1–25.

Tarski, A. Der Aussagenkalkül und die Topologie, *Fund. Math.* **31**, 103–134.

Tierney, M. (1972), Sheaf theory and the continuum hypothesis, in Lawvere (1972), pp. 13–42.

Troelstra, A.S. (ed.), (1973), *Metamathematical Investigation of Intuitionistic Arithmetic and Analysis*, Lecture Notes in Mathematics 344, Springer-Verlag, New York.

Troelstra, A.S. (1977), Aspects of constructive mathematics, in Barwise (1977), pp. 973–1052.

Troelstra, A.S. (1978), A. Heyting on the formalization of intuitionistic mathematics, in Bertin et al. (1978), pp. 153–175.

Troelstra, A.S. (1981), Arend Heyting and his contribution to intuitionism, *Nieuw Arch. Wisk.* **29**, 3rd series, No. 1, 1–23.

Troelstra, A.S. and van Dalen, D. (eds.), (1982), *The L.E.J. Brouwer Centenary Symposium*, Studies in Logic and the Foundations of Mathematics, Vol. 110, North-Holland, Amsterdam.

van Heijenoort, J. (ed.), (1967), *From Frege to Gödel. A Source Book in Mathematical Logic, 1879–1931*, Harvard University Press, Cambridge, MA.

Whitehead, A.N. and Russell, B. (1925), *Principia Mathematica*, "Second edition, vol. I, Cambridge University Press, 1925; "Second edition, vol. II," Cambridge University Press, 1927; "Second edition, vol. III," Cambridge University Press, 1927. Cambridge, UK.

The Writing of
Introduction to Metamathematics

Stephen C. Kleene

It was suggested that I talk here on the writing of my book *Introduction to Metamathematics* (IM).

In the summer of 1936 I drew up an outline for a graduate course on the foundations of mathematics, the first to be given at the University of Wisconsin, Madison. It ran in the fall semesters of 1936–37, 1938–39, and 1940–41, as well as more recently.

In the fall of 1939, Saunders MacLane suggested to Rosser and me that we jointly write a Carus monograph on foundations. But instead of doing that (the Carus monographs are rather compact books, tending to be of less than 200 pages), each of us separately wrote a book of over 500 pages. My *Introduction to Metamathematics*, was published in 1952. The ninth reprint by the Dutch publishers was in 1988.[1] According to incomplete records, about 17,500 copies were sold through 1986, not counting sales of a reprint in Taiwan and of two in Japan, of two printings of the Russian translation (the first consisting of 8000 copies), of one of the Spanish translation, and of one of the Chinese translation (in two volumes).[2] I must leave to others the assessment of the role the book may have played in the teaching of mathematical logic and foundations over the years.

I calculated that all of my spare time for $7\frac{1}{2}$ years went into the composition of the book. But the earlier preparation of my logic course was determining for the first ten chapters, which essentially followed that course. Subsequent chapters largely contain material used in seminars given in the spring semesters following the course beginning with 1938–39, except for the last two chapters (XIV and XV).

So what went into the course, the seminars, and the book? Mostly topics that stood out in the landscape of mathematical logic for me as an observer in the 1930s.

In 1983, I wrote, "Gödel's paper *1931* was undoubtedly the most exciting and the most cited article in mathematical logic and foundations to appear in the first 30 years of the century."[3] I learned of that paper, shortly after its publication, while I was a graduate student at Princeton. In subsequent research, I used Gödel's method (which I learned from it) of numbering the formal objects, and the Herbrand–Gödel notion of "general recursive functions" (from Gödel *1934*), in giving another version of Gödel's first *1931* incompleteness theorem in *1936*, and more simply in *1943* (abstract published in 1940) in connection with my arithmetical hierarchy (given independently by Mostowski in *1947*). I had originally intended to publish the details on my arithmetical hierarchy (§57 in the book) only through the book, hoping thereby to promote its sales. But I decided at the beginning of 1942 to publish them separately, when I realized that the publication of the book was far off. The writing of it had to be set aside during my military service in 1942–1945, while the uncompleted manuscript reposed in Saunders MacLane's office in Widener Library at Harvard.

What came to my mind in 1936 for my new course on the foundations mathematics? Gödel's *1931* results applied to Hilbert's uncompleted program for vindicating classical mathematics (which had been shaken by the paradoxes of set theory) by embodying a suitable portion of it in a formal system and proving that system consistent by safe ("finitary") methods in a new mathematical discipline to be called "proof theory" or "metamathematics." It struck me that the most exciting developments in foundations then centered around this program, and results such as Gödel's which came out of contemplating it. So I drew up a syllabus for the course which would give the student the whole broad picture of Hilbert's program (with the context in which it arose and with comparisons with the other outstanding foundational approaches), and which would allow one to present Gödel's theorems conveying a full understanding of them.

Thus I had to present the idea of a formal system, and train the student intensively in working with one. As I remarked in the preface of the book, the simplest formal system for the purpose of exhibiting such results as Gödel's is one for elementary number theory (based on first-order logic).

I put into the course (and book) three introductory chapters to set the enterprise in its broad historical context, and to present necessary fundamental concepts.

Then there followed (in Chapters IV–VIII) my specimen of a formal system, studied in stages. I think students of logic, confronted with the axioms and rules of inference of a formal system, had to rack their brains rather hard to construct proofs of formal theorems in it. The deduction theorem of the first-order predicate calculus (first proved by Herbrand in *1930*) provides great assistance in proving implications. It was emphasized in another context in Church's logic course in the fall of 1931–32. I used a similar treatment of disjunctions (proof by cases) in my *1934* paper. In brief, it is an easy step from the deduction theorem to a similar treatment of proofs of formulas involving each of the logical constants. Thus (originally in my 1936–37 course) I came to my treatment of logic in IM, §23 via a set of derived rules for the introduction and elimination of logical symbols. (One also finds versions of it in Jaśkowski *1934*, Gentzen *1934–35*, and Bernays, *1936*.) I followed von Neumann *1927* when I was treating logic as a subsystem of formal number theory, in using axiom schemata instead of propositional and predicate variables with formal substitution rules for them.

I arrived at the formulation and basic discussion of Gödel's two famous *1931* incompleteness theorems at the end of Chapter VIII. In the next two chapters (IX and X) I gave a treatment of primitive recursive functions (what Gödel in *1931* called simply "recursive functions") and completed the proof of his first incompleteness theorem.

I continued in seminars, and thence in the book, with chapters on general recursive functions (Gödel *1934*, adapting a suggestion of Herbrand, and Kleene *1936*), partial recursive functions (Kleene *1938*, discovered late in 1936), and computable functions (Turing *1936–37* and Post *1936*).

My treatment of computable functions, considerably reworked from Turing's account and in some respects closer to Post *1936*, was first introduced in my seminar in the spring of 1941. (That seminar was honored by the presence of Richard Brauer.) Turing applied his machines primarily to compute the dual expansions of real numbers x ($0 \leq x \leq 1$), the successive digits being printed *ad infinitum* on alternate squares of a one-way infinite tape, while the intervening squares were reserved for temporary notes used as scratch work in the continuing computation. (Many of Turing's technical details were incorrect as given, and a person who wishes to follow his treatment in detail will profit from the critique in the appendix to Post *1947*.) I respected and took over Turing's brilliant conception of the kinds of operations a human, or physical computer can perform, and thus the basic mode of operation of his machines, But, continuing from my Chapters IX, XI, and XII, I wanted rather to consider the computability of number-theoretic functions. I chose to use a machine of his sort to compute a given such function by supplying it with a tape on which is printed an argument (or n-tuple of arguments) of the function, and asking that the machine eventually come to a stop with the corresponding function value printed on the tape following the argument(s). A natural number n, as argument or value, I represented by $n + 1$ tallies printed on consecutive squares, with blank squares separating the representations of two numbers.

In his §10, Turing suggests that possibly the simplest way to use a machine of his to compute, e.g., a function $\phi(n)$ of "an integral variable," n would be to let it compute the sequence of 0s and 1s with the number of 1s between the nth and $(n + 1)$th 0 being the value of $\phi(n)$.[4] That, in my opinion, would certainly be more cumbersome to work with than my method. Moreover, that method breaks down when one seeks to apply it to the partial recursive functions which I introduced in *1938* (and §63). Wang (*1974*, p. 84) says, "Gödel points out that the precise notion of mechanical procedures is brought out clearly by Turing machines producing partial rather than general recursive functions. In other words, the intuitive notion does not require that a mechanical procedure should always terminate or succeed. A sometimes unsuccessful procedure, if sharply defined, still is a procedure, i.e., a well determined manner of proceeding. Hence we have an excellent example here of a concept which did not appear sharp to us but has become so as a result of a careful reflection. The resulting definition of the concept of mechanical by the sharp concept of 'performable by a Turing machine' is both correct and unique. Unlike the more complex concept of always-terminating mechanical procedures, the unqualified concept, seen clearly now, has the same meaning for the intuitionists as for the classicists. Moreover, it is absolutely impossible that anybody who understands the question and knows Turing's definition should decide for a different concept."

I remember that, when in the spring of 1940 I used the words "partial recursive function" in a conversation with Gödel, he immediately asked, "What is a partial recursive function?". Church in *1936* had avoided the concept of "partial recursive function" by using "potentially recursive function."

Having started out to build the course (and the first ten chapters of the book) toward getting Gödel's incompleteness theorems, it was essentially irresistible to proceed to these next three chapters (XI, XII, and XIII). There had been such exciting developments in the foundational research at Princeton (involving Church, Gödel, Rosser, and myself) just before my coming to Wisconsin in the fall of 1935, culminating in Church's thesis (proposed in 1934 and 1935 and published in 1936) and its applications, and in Turing's *1936–7* work, and Post's *1936*, of each of which I learned a bit later! For the history of these developments, see Kleene (*1981*) and Davis (*1982*).

In brief, when I was invited in 1936 to teach a graduate course in the foundations of mathematics, and later (first in 1939) to extend it by a seminar, these were the topics which seemed to me the most exciting. Naturally, I was influenced by where my own research interests lay. There was no existing *connected* treatment of them in an English text book that I could just follow. (Quite a bit of the material was available in German in Fraenkel *1928*, Hilbert and Ackermann *1928*, Heyting *1934*, and Hilbert and Bernays *1934* and *1939*.) So I designed the course (giving my students dittoed notes), expanded it in seminars, and subsequently wrote it all into my own textbook.

Incidentally, the symmetric form of Gödel's theorem (§61), which I discovered in 1946–47, was in the manuscript of my book when I showed it to Mostowski on his visit to Madison in April 1949, on whose urging I published it separately in *1950*. Kreisel wrote me on 4 October 1983 that Gödel on several occasions remarked on this symmetric form as being a very significant improvement of his incompleteness results. Kreisel continued, "The reasons are so patently clear that I did not bother to ask him to elaborate." I did not ask Kreisel to elaborate. But I will attempt now to do so myself.

Fundamental requirements on a formal system are as follows:

(1) It should be effectively recognizable what linguistic objects (formulas) of the system express certain well-defined propositions of mathematics— those of a domain of intuitive mathematics that we are choosing the formal

system to include a formalization of. If this domain includes some elementary number theory, then by Church's *1936* thesis discussed in §62, to any well-defined one-place number-theoretic predicate (propositional function of one natural number variable x) of this domain, the Gödel number of the formula expressing the proposition taken as its value for a given natural number as argument (i.e., as value of x) will be a general recursive function of the argument x. We are assuming Gödel numbers assigned to symbols, strings of symbols, and strings of strings of symbols in the standard way.

(2) It should be possible to check effectively without fail whether a string of formulas we have before us constitutes a proof in the system. Then, by Church's thesis, to be the Gödel number of a proof is a general recursive predicate. There is a primitive recursive function which, applied to the Gödel number of a proof, gives the Gödel number of the formula proved by it (its last formula).

(3) And of course, for any propositions to which the qualities of being true or false apply in a clear way, we want a proof of the formula asserting one of those propositions to exist only when the proposition is true. For my symmetric form of Gödel's theorem, I use only as much of this as is assured by simple consistency (after Rosser *1936* and my *1943*, top p. 64).

In the *1931* version of Gödel's first incompleteness theorem, the formal systems to which it applied were "*Principia Mathematica* (Whitehead and Russell *1910–12*, second edition *1925, 1927*) and related systems." It was far from clear then that the theorem would apply to all formal systems satisfying in reasonable measure the aforesaid requirements (inclusive of systems quite remote in their details from Gödel's examples), except ones so weak (unrobust) as to be of slight interest.

That the theorem applies to all conceivable simply consistent slightly robust systems is given by my symmetric form of Gödel's theorem. For the robustness, what I ask is that they include a formalization of the small piece of elementary number theory which I now describe.

I use two recursively enumerable sets $C_0 = \check{x}(Ey)W_0(x, y)$ and $C_1 = \check{x}(Ey)W_1(x, y)$ (the variables range over the natural numbers 0, 1, 2, ...), where $W_0(x, y)$ and $W_1(x, y)$ are two particular primitive recursive predicates (defined on p. 308). For each of $i = 0$ or 1, and any given x such that $x \in C_i$ (i.e., such that $(Ey)W_i(x, y)$), a proof of that fact in intuitive elementary number theory exists, resting simply on our ability (because W_i is primitive recursive) to verifty for the given x and a suitable y that $W_i(x, y)$ is true. Furthermore, an easy argument in intuitive elementary number theory shows that my two sets C_0 and C_1 are disjoint.

For a formal system to come under my symmetric form of Gödel's theorem, it should formalize the foregoing, thus:

(a) For each of $i = 0$ and 1 and each x, it shall have a formula (which we can find effectively from the i and x) expressing the proposition $x \in C_i$ (i.e.,

$(Ey)W_i(x, y))$, and likewise one expressing $x \notin C_i$. I shall denote these formulas by "$\exists y W_i(x, y)$" and "$\neg \exists y W_i(x, y)$," respectively, avoiding placing any restriction on the actual symbolism of the system; and for simple consistency I merely understand that for no i and x are both of those formulas provable.

(b) Corresponding to the intuitive provability of $x \in C_i$ whenever true, in it for each of $i = 0$ and each x:

(∗)   If $x \in C_i$, then $\exists y W_i(x, y)$ is provable.

(c) Corresponding to the intuitive proof of the disjointness of C_0 and C_1, in it for each x:

(∗∗)
    If $\exists y W_0(x, y)$ is provable, then $\neg \exists y W_1(x, y)$ is provable.

    If $\exists y W_1(x, y)$ is provable, then $\neg \exists y W_0(x, y)$ is provable.

My symmetric form of Gödel's theorem shows that, for each formal system which encompasses this small piece of number theory and is simply consistent (no matter what novelties of symbolism, or of methods of proof, it may have), we can find a number f such that the formula $\neg \exists y W_0(f, y)$ is true but unprovable while $\exists y W_0(f, y)$ is also unprovable (and symmetrically, a number g such that $\neg \exists y W_1(g, y)$ is true while it and $\exists y W_1(g, y)$ are both unprovable).

Thus, for each such formal system, we have as a formally undecidable proposition $(Ey)W_0(f, y)$ a respective value of one preassigned predicate $(Ey)W_0(x, y)$, a result which Gödel did not have in 1931. Just the theory of that particular predicate provides inexhaustible scope for mathematical ingenuity.[5]

The symmetric form of Gödel's theorem, as I stated it in abstract terms in 1950, says simply that to any two disjoint recursively enumerable sets D_0 and D_1 with $D_0 \supset C_0$ and $D_1 \supset C_1$, a number f can be found such that $f \notin D_0 \cup D_1$. Applying this by taking D_0 to be the x's such that $\exists y W_0(x, y)$ is provable (so by (∗), $D_0 \supset C_0$) and D_1 to be the x's such that $\neg \exists y W_0(x, y)$ is provable (so by the simple consistency, D_1 is disjoint from D_0; and by (∗) and (∗∗), $D_1 \supset C_1$), the f we get does the job.[6]

Chapter XIV picked up some standard topics of mathematical logic which one would recognize in the 1930s as important and which had not found a place in Chapters I–XIII. Those earlier chapters provided background material, using which these topics could be treated compactly. Among these topics were some outstanding model-theoretic results (Gödel's 1930 completeness theorem, including the theorem of Löwenheim 1915 and Skolem 1920 extended to include "compactness," in §72; and Skolem's 1922–23 paradox, and his 1933, 1934 nonstandard arithmetics, in §75), as well as some more topics in proof theory (metamathematics). After all, I had to live up to the title of the book; and one of the jewels of proof theory was eliminability theory after Hilbert and Bernays 1934; so I put in §74. Even Gödel's completeness theorem for the predicate calculus (§72) has a proof-theoretic version after Hilbert and Bernays 1939 (IM, Theorem 36) and a relation to my arithmetical hierarchy (IM, Theorems 35 and 38).

As to the final Chapter XV, I first became aware of Gentzen's elegant *1934–35* paper in 1947, and used it to expound some results on consistency proofs in §79. And I concluded with §§80–82 on intuitionistic systems, the last giving an exposition of research initiated by me in 1941, and published by me in *1945* and David Nelson (in his Ph.D. thesis written under my direction) in *1947*. The first pages of Kleene (*1973*) give the history. This research established a connection between intuitionism and the subjects of general and partial recursive functions and Church's thesis, which were in the main line of the book (Chapters XI and XII). I remember Tarski telling me that he found this an interesting connection, which he had not thought of searching for himself. I was not under the spell of Brouwer, except to the extent that, when something can be done intuitionistically (not just classically), I feel it has more concrete content. Indeed, §82 emphasizes this. I had many interesting conversations with Brouwer from 1948 on, and felt he was sympathetic to my work. Many logicians not identified as intuitionists have paid attention to intuitionism (e.g., Kolmogorov (*1932*); Gödel *1932, 1932–33*; Gentzen *1934–35*; Jaśkowski *1936*; Gentzen *1936*, p. 532 and Bernays (cf. IM, p. 495)). I had from the beginning made a point of identifying (at very little cost of space) which results hold for the intuitionistic versions of the formal systems being considered.

I have been asked why I did not include much model theory, or go further into set theory. Little of what is presented now in an introductory model theory course was known in 1952 when my book was published. As remarked in Chapter XIV, I did include some outstanding results. As to set theory, Gödel's work on the consistency of the continuum hypothesis first came out in *1938* (and Cohen's completion of an independence proof for it in *1963–64*), which was after the basic choices had been made as to the direction my course, seminars and book would take. Unlike the selections from model theory treated in Chapter XIV, this topic (involving details of the development of set theory formulated in logical symbolism, rather than just of logic and number theory) could not have been added without taking considerable space.

I intended to make the book essentially self contained, so that a reader with typical undergraduate-mathematics-major preparation could follow the story I was telling without having to supplement it by outside sources. At the same time I endeavored to be generous with references to the literature, both to give credit as due, and also (sometimes stating results) with the idea of encouraging the reader to enlarge his knowledge beyond what he could learn from the book.

I hoped all the topics I chose to work into the book would be of lasting significance. As I treated them, they made too bulky a manuscript to leave room for any other massive development.

Had I made the manuscript of the book any bigger, it might not have been published. *Proof*: The manuscript was accepted in principle on 4 April 1950 for publication by D. Van Nostrand Co. But on 31 August 1950, they wrote me backing out on the basis of the estimate by their printer that it would be

600 pages and would have to sell several thousand copies in the next few years to recoup the cost of printing. I replied that, by my calculation from a line count of my typescript, it would be 550 pages.[7] I then persuaded the North-Holland Publishing Company (with whom I had been in personal contact in Amsterdam in the spring of 1950 about printing the *Journal of Symbolic Logic*) to share with P. Noordhoff Ltd. the risk of printing my book, with Van Nostrand agreeing to bring out an American. edition from sheets printed in the Netherlands.

Notes

1. There has been no revision—only a few equal-space changes, two notes added in open spaces at ends of chapters, and the updating of eleven references which in 1952 were "to appear," as is indicated on p. vi of the sixth (1971) and later reprints.

2. On 31 December 1984 Moh ShawKwei, who translated IM into Chinese in 1965, wrote me "Now at last the Chinese translation of your work *Introduction to Metamathematics* comes off the press—only the first half of it yet. The other half, I hope, will be pressed in about a year."

3. Italicized dates, as here "*1931*," constitute references to the Bibliography of *Introduction to Metamathematics*, or (mainly for dates after 1952) to its Supplement at the end of the present article. IM was the first work, in any area of scholarship with which I am acquainted, to use dates (years) for the references.

4. This covers the case Turing's "integral variable" n ranges over the positive integers. If it ranges over the natural numbers, the number of 1's preceding the first 0 shall be the value of $\phi(0)$.

5. In *1943* and IM, §60, I used as the preassigned predicate $(y)\overline{T}_1(x, x, y)$ with the assumption of ω-consistency; and in *1936* $(y)(Ez)T_1(x, y, z)$ with a more complicated consistency assumption.

6. That D_0 and D_1 are recursively enumerable follows from the above fundamental requirements (1) and (2) using Church's thesis.

7. One may observe that, not counting the ten preliminary pages i–x with the preface and table of contents, it turned out to be exactly 550 pages, with the last page of the text, 516, filled to the last line.

Supplement to the Bibliography in IM

Cohen, P.J. (1963–4), The independence of the continuum hypothesis, and ibid., II. *Proc. Nat. Acad. Sci.* **50**, 1143–1148 and **51**, 105–110.

Davis, M. (1982), Why Gödel didn't have Church's thesis, *Inform. and Control*, **54**, 3–24.

Kleene, S.C. (1973), *Realizability: A retrospective survey*. Cambridge Summer School in Mathematical Logic, 1971, in A.R.D. Mathias and H. Rogers (eds.), Lecture Notes in Mathematics, No. 337, Springer-Verlag, Berlin, pp. 95–112.

Kleene, S.C. (1981), Origins of recursive function theory. *Ann. Hist. Comput.* **3**, 52–67. For six corrections, see Davis (1982), footnotes 10 and 12.

Kolmogorov, A.N. (1932), Zur Deutung der intuitionistischen Logik, *Math. Z.* **35**, 58–65.

Wang, H. (1974), *From Mathematics to Philosophy*, Routledge and Kegan Paul, London and Humanities Press, New York, xiv + 428 pp.

In Memoriam: Haskell Brooks Curry

Jonathan P. Seldin

Haskell Brooks Curry died on 1 September 1982 at the age of 81. He will be remembered as one of the founders of mathematical logic in the United States. In particular, he did much to develop combinatory logic, with which his name will always be associated.

He was born on 12 September 1900, at Millis, Massachusetts. His father was president of the School of Expression in Boston, which is now known as Curry College. He graduated from high school in 1916 and received his A. B. in mathematics from Harvard University in 1920.

He began his graduate work in electrical engineering at MIT in a program that involved working half-time at the General Electric Company, but he grew more interested in pure science than in engineering, and in 1922 he returned to Harvard to study physics. He received his A.M. in physics from Harvard in 1924. Then, in line with his increasing interest in foundations, he went on to further study in mathematics, his undergraduate major.

He continued to study mathematics at Harvard until 1927; during this period his interest in mathematical logic grew more and more important. This interest, which had begun as early as 1922, originally involved reading on the side. He had, in fact, been advised by several faculty members from Harvard and elsewhere, including Norbert Wiener, to stay away from logic. But during 1926–27 he had an idea that caused Wiener and others to change their minds and encourage him to pursue logic. His idea had to do with the rule of substitution in the first chapter of Whitehead and Russell, (1910–1913); this rule and *modus ponens*, are the only rules given there for the propositional calculus. Curry noticed that the rule of substitution is considerably more complicated than *modus ponens*, and he wanted to break it down into simpler rules. His idea was to use what he later called combinators for this purpose.

As a result of this idea and the encouragement it generated, he decided to change the subject of his dissertation to logic. In 1927–28, he went to Princeton to assume an instructorship, where he pursued this idea. In November 1927, he found in the Library at Princeton (Schönfinkel, 1924), which anticipated his notion of combinator. Unfortunately, Schönfinkel was by then in a mental hospital and was unlikely to do any more scientific work.

Fortunately, Paul Bernays at Göttingen was familiar with his work, and so Curry decided to go there.

Before leaving the USA, he married on 3 July 1928, Mary Virginia Wheatley of Hurlock Maryland, a former student at the School of Expression. They were to have two children, Anne and Robert, and seven grand-children.

At Göttingen, Curry immediately set to work on his dissertation (Curry, 1930), in which he set up a system of what we now call pure combinatory logic. (The dissertation contained a proof that the system was what we now call combinatorially complete, a proof of consistency, and a finite axiomatization of the extensionality rule.) It was completed within one academic year, and Curry was examined on 24 July 1929. Although he had done most of his work with Bernays, the referee was David Hilbert. Some of his memories of the examination are recorded in Reid (1970, p. 190).

In September 1929, the Currys moved to State College, Pennsylvania, where Curry took up a position on the faculty of the Pennsylvania State College. For the next five years he extended the system of his dissertation in a series of papers culminating in Curry (1934a), in which he was able to interpret the first-order predicate calculus. He had also begun to think about strengthening the system to one of type-free logic in which set theory could be interpreted; see the abstract in Curry (1934b). In this work he was aided by the opportunity to spend the year 1931–32 at the University of Chicago as a National Research Council Fellow.

This line of research was paralleled by the work of Church and his students, Kleene and Rosser, on type-free logic based on λ-calculus; see, for example, Church (1932). All of this work was brought to a halt in 1934 by Kleene and Rosser, who found a contradiction that could be derived in the system of Curry (1934a); see Kleene and Rosser (1935). After this Church and his students gave up the idea of using a system of this sort as a basis for ordinary systems of mathematical logic. But as explained in Seldin (1980), Curry was prepared for the possibility that his system might be inconsistent and had some ideas for dealing with this without giving up his original program. He spent the rest of the 1930s analyzing the contradiction discovered by Kleene and Rosser and developing a program for a new system of combinatory logic that would be consistent and still strong enough to serve as a basis for systems of quantification logic in the ordinary sense. During this period he became acquainted with the work of Gentzen on consistency proofs.

Meanwhile, Curry became a founding member of the Association for Symbolic Logic. He was Vice-President in 1936–37 and President in 1938–40.

In 1938–39 he spent a year in residence at the Institute for Advanced Study at Princeton.

At the end of the 1930s, he was asked to present his views on the nature of mathematics to the International Congress for the Unity of Science scheduled for Cambridge, Massachusetts, September 1939. The result was (Curry, 1939), the first of a long series of papers in which he developed the philosophy of mathematics that he called formalism. (The basic ideas for this philosophy were already present in his earliest published work; see Seldin, (1980, p. 11).)

By 1941 and early 1942, he had worked out a specific program of research in combinatory logic. applying Gentzen's techniques to obtain consistency proofs of a number of systems. He published this program as Curry (1942b). But when the United States entered World War II, he decided to put logic aside and devote his energies to the war effort. In May 1942, he left Penn State and went to the Frankford Arsenal, where he remained until January 1944. He then went to the Applied Physics Laboratory at Johns Hopkins University and remained there until, in March 1945, he moved on to the Ballistics Research Laboratories at the Aberdeen Proving Ground. His work on the problem of fire control brought him into contact with the ENIAC, an early computer designed to compute trajectories of artillary shells and other projectiles. For his last three months at the Aberdeen Proving Ground, he was Chief of the Theory Section of the Computing Laboratory. This work led to a series of publications on computing for about a decade after the war, (see Hindley and Seldin (1980, pp. xiii–xx)).

In September 1946, Curry returned to Penn State. He wanted to continue work in computing, but he was unable to interest the Penn State administration in venturing into this new field. Thus, he turned back to logic.

When he received an invitation to give a series of lectures at the University of Notre Dame, Indiana, he decided to devote the lectures to Gentzen's techniques. The resulting lectures were written up and published as Curry (1950).

During the summer of 1948, he went back to Europe to attend the Tenth International Congress of Philosophy. There he was asked to write a short book of less than 100 pages on combinatory logic for a series on logic to be put out by the newly formed North-Holland Publishing Company. The people who asked him had no idea how much unpublished material there was on the subject and of how much work needed to be finished in order to write a proper presentation. So he sent them *Outlines of a Formalist Philosophy of Mathematics* (Curry, 1951) instead. But he kept thinking about a book on combinatory logic, and finally decided to ask Robert Feys of the University of Louvain, Belgium, to collaborate with him. In 1950 Curry obtained a Fulbright grant to spend the year 1950–51 there working on this book. While there, he also wrote *Leçons de Logique Algebrique*. The book on combinatory logic was not finished until 1956; it appeared as *Combinatory Logic*, volume I. It contained a considerable amount of new research, and among its results were a thorough study of the Church–Rosser Theorem for λ-calculus (including an extension to $\lambda\beta\eta$-reduction), standardization theorems for both $\lambda\beta$- and $\lambda\beta\eta$-reduction, and a proof of the normal form theorem for functionality. (The theory of functionality, which is also known as simple type assignment, is similar to Church's type theory; see Church (1940).)

After *Combinatory Logic*, volume I, was published, Curry again turned his attention to Gentzen's proof theory, producing the book *Foundations of Mathematical Logic* (Curry, 1963). Part of his reason for doing this was that he wanted to use the Gentzen techniques in connection with a proposed second volume on combinatory logic. This second volume would develop systems of combinatory logic which would be an adequate foundation for type-free higher order logic. Two papers (Curry, 1960, 1961) began the development of the theory of restricted generality along the lines already outlined in Curry (1942b). (Restricted generality concerns quantifiers of the form $(\forall x \in A)B(x)$; Curry had observed that most quantifications in ordinary mathematics are over restricted ranges.) He was interested in this theory partly because his original reaction to the paradox of Kleene and Rosser had been that the problem was in the postulates for the universal quantifier, and this interest remained even after he showed (Curry, 1942a) that a contradiction could be derived from the postulates for implication alone.

In 1960, Curry became Evan Pugh Research Professor at Penn State, and was thus relieved of undergraduate teaching duties. This undoubtedly helped him finish *Foundations of Mathematical Logic* (Curry, 1963) sooner than he otherwise would have, and enabled him to turn his full attention to the second volume on combinatory logic in 1964. Robert Feys had died in 1961, and the amount of.new research needed for this book was large; as a result, he invited Roger Hindley to join him on the project in 1965 and J.P. Seldin in 1968. But this is getting ahead of the story.

In 1966, he retired from Penn State after 37 years and took the position of Professor of Logic, History of Logic, and Philosophy of Science and also director of the Instituut voor Grondslagenonderzoek en Philosophie der Exacte Wetenschappen (Institute for Foundation Studies and Philosophy of the Exact Sciences) at the University of Amsterdam. He stayed at Amsterdam for four years until he retired again in 1970.

The year 1970 was also the year the manuscript for *Combinatory Logic*, volume II was finished; it appeared in 1972. In addition to revising parts of volume I (Curry and Feys, 1958), it contains a chapter on representing recursive functions, one on restricted generality, one on implication and the universal quantifier, and one on type theory.

In 1971–72 Curry was Visiting Mellon Professor of Mathematics at the University of Pittsburgh. Otherwise, he lived at State College, Pennsylvania (the town where Penn State is located), where he continued his research (producing five papers in the 1970s) and his reviewing work (he was a regular reviewer for the *Zentralblatt für Mathematik und ihre Grenzgebiete* 1931–1939, *Mathematical Reviews* 1940–1979, and *Computing Reviews* 1965–1979; he also wrote several reviews for the *Journal of Symbolic Logic* and other publications).

Even in 1978, when he was 78 years old, he was still working on a research problem: that of finding a reduction relation (called "β-strong reduction") that would go with combinatory β-equality and correspond to $\lambda\beta$-reduction. But in the summer of 1979 he suffered a stroke and had to give up all scientific work for more than a year. Nevertheless, he recovered enough to resume work on the β-strong reduction problem in 1981. He was still working on this in the spring of 1982. A progress report of this work was presented by J.P. Seldin to Logic Colloquium '82, Florence, Italy in August 1982; see Curry et al. (1984). In July Curry's health began to deteriorate, and although he was able to attend the Association for Computing Machinery Symposium on LISP and Functional Programming at Pittsburgh, Pennsylvania, 15–18 August 1982 (where he and Alonzo Church were jointly honored at a banquet), he was by that time very ill. He died of a stroke on 1 September 1982.

Curry's mathematical career spanned close to 60 years, and he was principally responsible for the development of a new field, combinatory logic. In the beginning, this was seen mainly as a tool in the foundations of mathematics. When this avenue appeared blocked by paradoxes (to just about everybody except Curry and F.B. Fitch), interest declined for awhile. But by the 1970s, Curry was gratified to see interest grow again and become greater than before. His studies in "pure" combinatory logic have been taken up and developed by Barendregt and his group, and "applied" combinatory logics are currently of interest as tools in computer science.

In fact, there is more interest in Curry's work among computer scientists today than there is among logicians. This is, in my opinion, only partly because both λ-calculus and combinatory logic (in both their typed and untyped versions) are useful as prototypes for many of the new languages now being designed. It is also partly a result of the fact that Curry came to logic from applied mathematics. Most mathematical logicians come to the subject from pure mathematics or from philosophy, and as a result bring with them strong opinions as to what is logically "correct." Curry did not have most of these strong opinions, but took a strongly pragmatic attitude towards the acceptance (or rejection) of a formal system (which he always considered to be dependent on a specific purpose). This lack of strong attitudes and pragmatic approach seemed strange to most other logicians. In addition, many logicians had trouble understanding why he was so interested in matters that seemed too technical to bother with. But now that computers have grown powerful enough to do meaningful symbolic calculations, it has become apparent that these "technical" matters are extremely important for practical implementations. Furthermore, most computer scientists react to their subject the same way other applied mathematicians do, and thus find Curry's pragmatic attitude essentially the same as their own.

Curry had the following Ph.D. students (the dates in parentheses are the dates they finished their dissertations): Edward J. Cogan (1955), Kenneth L. Loewen (1962), Bruce Lercher (1963), Luis E. Sanchis (1963), Jonathan P. Seldin (1968), and Martin W. Bunder (1969). John A. Lever also wrote a masters thesis with him in 1977.

In addition to his mathematical work, Curry was widely known for his interest in ornithology. He considered it a hobby and described it as bird-watching, but he pursued it with some seriousness. Among his mathematical notes (which are filed by date rather than by subject) can be found several reports of bird sightings. The hobby has also led to some interesting incidents (Gaber, 1949):

> In Norway in 1928, while unintentionally trespassing upon the king's estate, he narrowly escaped being chewed up by two of the palace dogs and was saved from them by a tall handsome man. Ten years later, upon seeing the pictures in the papers of the members of a Norwegian mission to this country, he was quite surprised to learn that his rescuer was none other than the Crown Prince Olaf of Norway.
>
> His hobby has even afforded him the opportunity of having a laugh at the expense of one of his colleagues. Having an exam scheduled for one of his classes one afternoon, he arranged to have the test papers taken to the room where the test was to be given while he went out for a few hours in search of birds. He arrived back from his walk too late to stop off at his office and leave his binoculars there, so he went directly to where the test was being given. When he walked into the room with his binoculars slung over his shoulder, one of the more excitable professors on hand spotted him and exclaimed, "My God, Curry, you're not going to use those things to check on the students!" "Why, certainly," replied Curry. He immediately stationed himself in the rear of the room and proceeded to scan the room with his glasses. On subsequent exams when the excitable professor was to be present, he made it a point to carry his binoculars with him.

It is possible that some people would conclude on meeting Curry for the first time that he was somewhat distant. But those of us who knew him better knew otherwise. This apparent aloofness would be manifest only when his mind was on a mathematical problem, for like all good mathematicians he had great powers of intense concentration. As the end of the biographical sketch in Hindley and Seldin (1980) says of him and his wife:

> Everybody who know the Currys is aware of how friendly and helpful they always are. Haskell has always done more for colleagues and students than be a source of important ideas (although, of course, his ideas have been of tremendous importance). He has also always been willing to listen to anybody who wanted to talk to him, to discuss their ideas, and to give whatever encouragement he could. (Surely many of us have heard stories of his taking time to help a student having trouble in an elementary course taught by somebody else.) His office door has always been open. And this has undoubtedly been an important contribution to the enthusiasm of many of those of us working in combinatory logic. Also well known wherever the Currys have lived has been the hospitality they have both shown. There [were] always many parties and other, less formal gatherings, and we conjecture that Virginia's cooking has also played a role in the growth of interest in combinatory logic.

Acknowledgment

I would like to thank Roger Hindley for his helpful criticisms and suggestions.

References

Church, A. (1932), A set of postulates for the foundation of logic, *Ann. of Math.* (2) 33, 346–366.

Church, A. (1940), A formulation of the simple theory of types, *J. Symbolic Logic* **5**, 56–68.

Curry, H.B.[1] (1930), Grundlagen de kombinatorischen Logik (Inauguraldissertation), *Amer. J. Math.* **52**, 509–536, 789–834.

Curry, H.B. (1934a), Some properties of equality and implication in combinatory logic, *Ann. of Math.* (2) **35**, 849–860.

Curry, H.B. (1934b), Foundations of the theory of abstract sets from the standpoint of combinatory logic (Abstract), *Bull. Amer. Math. Soc.* **40**, 654.

Curry, H.B. (1939), Remarks on the definition and nature of mathematics, *J. Unified Science* **9**, 164–169.

Curry, H.B. (1942a), The inconsistency of certain formal logics, *J. Symbolic Logic* **7**, 115–117.

Curry, H.B. (1942b), Some advances in the combinatory theory of quantification, *Proc. Nat. Acad. Sci. U.S.A.* **28**, 564–569.

Curry, H.B. (1950), *A Theory of Formal Deducibility*, Notre Dame Math. Lectures No. 6, Notre Dame, Indiana.

Curry, H.B. (1951), *Outlines of a Formalist Philosophy of Mathematics*, North-Holland, Amsterdam.

Curry, H.B. (1952), *Leçons de Logique Algebrique*, Gauthier-Villars, Paris and Nauwelaerts, Louvain.

Curry, H.B. (1960), The deduction theorem in the combinatory theory of restricted generality, *Logique et Analyse* **3**, 15–39.

Curry, H.B. (1961), Basic verifiability in the combinatory theory of restricted generality, In *Essays on the Foundations of Mathematics*, Magnes Press of Hebrew University, Jerusalem, pp. 165–189.

Curry, H.B. (1963), *Foundations of Mathematical Logic*, McGraw-Hill, New York.

Curry, H.B. and Feys, R. (1958), *Combinatory Logic*, Vol. I, North-Holland, Amsterdam.

Curry, H.B., Hindley, J.R., and Seldin, J.P. (1972), *Combinatory Logic*, Vol. II, North-Holland, Amsterdam.

Curry, H.B., Hindley, J.R., and Seldin, J.P. (1984), Beta strong reduction in combinatory logic: Preliminary report (Abstract), *J. Symbolic Logic* **49**, 688.

Gaber, A.M. (1949), Profile: Dr. Haskell B. Curry, typescript of 8 pages filed by Curry under the date 11 April 1949.

Hindley, J.R. and Seldin, J.P. (eds.) (1980), *To H.B. Curry: Essays on Combinatory Logic, Lambda Calculus and Formalism*, Academic Press, London.

Kleene, S.C. and Rosser, J.B. (1935), The inconsistency of certain formal logics, *Ann. of Math.* (2) **36**, 630–636.

Reid, C. (1970), *Hilbert*, Springer-Verlag, New York and Berlin.

Schönfinkel, M. (1924), Über die Bausteine der mathematischen Logik, *Math. Ann.* **92**, 305–316.

Seldin, J.P. (1980), Curry's program, in Hindley and Seldin, (1980), pp. 3–33.

Whitehead, A.N. and Russell, B. (1910-1913), *Principia Mathematica*, 3 vols., Cambridge University Press, Cambridge, UK.

---

[1] A complete list of Curry's publications can be found in Hindley and Seldin (1980), pp. xiii–xx. The works listed here are those cited in this paper.

The Work of J. Richard Büchi[1]

Dirk Siefkes[2]

Abstract. J. Richard Büchi has done influential work in mathematics, logic, and computer sciences. He is probably best known for using finite automata as combinatorial devices to obtain strong results on decidability and definability in monadic second order theories, and extending the method to infinite combinatorial tools. Many consider his way of describing computations in logical theories as seminal in the area of reduction types. With Jesse Wright, identifying automata with algebras he opened them to algebraic treatment. In a book which I edited after his death he deals with the subject, and with its generalization to tree automata and context-free languages, in a uniform way through semi-Thue systems, aiming for a mathematical theory of terms. Less recognized is his concept of "abstraction" for characterizing structures by their automorphism groups, which he considered basic for a theory of definability. An axiomatic theory of convexity which originated therefrom is partly published jointly with W. Fenton. Also partly published is joint work on formalizing computing and complexity on abstract data types with the present author. Unpublished is, and likely will be, his continued work on an algorithmic version of Gauss' theory of quadratic forms, which stemmed from his interest in Hilbert's 10th problem. Results in the existential theory of concatenation, which link the two areas, are published jointly with S. Senger. Saunders MacLane and I edited a volume of *Collected Works of Richard Büchi*.

Working with J. Richard Buchi has been strenuous, frustating, and beautifully rewarding. Our meetings always lasted through the night. We would start with a heated political discussion, or inquire into a biological question—he had read most of Darwin. Then we would head into mathematics. If he brought up the subject, he would be the teacher and I the student, and I would learn a lot. He never, however, told me what he knew. Rather, he would work on some problem, in this way trying to make me see the things he saw. When I

---

[1] Died 11 April 1984.

[2] I did this work with the help of travel grants from Purdue University and from the Deutsche Forschungsgemeinschaft. They enabled me to visit Purdue University twice after Richard Büchi's death to look through his notes. I also thank the Deutsche Forschungsgemeinschaft, the Stiftung Volkswagenwerk, and the Fulbright Commission for earlier travel grants to work with Richard Büchi.

brought up the subject, it was worse. He would ask questions which might sound silly; I would answer; he would ask more questions; I would realize that I had better answer carefully; but soon my answers would not fit, and my ideas would crumble away. Then we would start together afresh, and something new and beautiful would evolve, although it might take a long time.

This is the way he dealt with mathematics, or any other field. He could not talk, or hear, *about* a subject. He had to work *through* it, each time anew, to make it his own. (For this reason he quarrelled with so many people: after a conversation both parties would think that the other had learned from himself.) This unrelenting way of working made him one of the great mathematicians of our time; he never stopped short of creating a clear and complete picture.

He did not, however, get stuck in the details either. He gave new directions to many fields. (I learned of many of them only after his death when I looked through his notes.) He always investigated the history of the area, and was deeply involved with the philosophical, social, and personal problems of people in mathematics and in science in general.

Now you see the double squeeze I am in: I am supposed to write about Richard Büchi's work, although he himself would have hated that; and much of it I do not even know well. So instead of writing about subjects and results and giving a complete overview, let me show you some of the things he liked to work with, and just mention some others. This still will not do, but it is the best that came to my mind.

1. Trees, Algebras, Automata, and Languages

Richard Büchi liked trees. So do I. Thus let me start with the trees which played a role in much of his mathematics. The full binary tree has a root e and successors $w1$ and $w2$ for each node w:

It represents the set $\{1, 2\}^*$ of words over the alphabet $\{1, 2\}$ with empty word e, where words are written left-to-right. Or in another formulation: it represents the free algebra generated by the element e and the unary operations $\cap 1$ and $\cap 2$. Every algebra with the same generators is a quotient (homomorphic image) of this one. Every finite such algebra is also a finite automaton over the alphabet $\{1, 2\}$: nodes are states, connected by paths in the tree (input words).

For an example take a finite automaton which accepts the language (set) L of words with an even number of 2s:

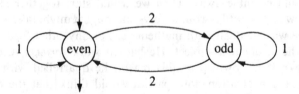

The two states tell whether the input has had an even or an odd number of 2s so far. Map the automaton into the tree by putting the initial state at the root and closing a path when you hit a state the second time:

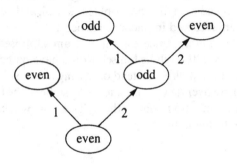

Identifying states gives back the automaton. So automata and unary algebras are really the same.

In the years 1952–1962 Richard Büchi and his coworkers in the Logic of Computer Group at the University of Michigan—especially Jesse Wright, Calvin Elgot, Jim Thatcher, Robert McNaughton, Michael Harrison, and Hao Wang—established and worked out these connections:

languages ↔ automata ↔ unary algebras ↔ grammars.

The relevant publications are nos. 10–12 in the bibliography, as well as the later no. 17. For years he worked on a book, no. 39, which contains all this material and its (by now well-known) generalization to the more-than-unary case:

term languages ↔ tree automata ↔ algebras ↔ tree grammars.

In both cases finite automata correspond to regular grammars (= semi-Thue systems). Richard Büchi goes back and forth between the unary and the general case to get the right notions of "regular," "automaton," etc. In the unfinished final part of the book, first for Polish notation and then for arbitrary terms, he generates, parses, and evaluates term languages through context-free grammars. He applies this approach to LRk-languages, and to an analysis of term algebras, aiming for a mathematical theory of mathematical notation. Most interesting seems to me the result that ground term rewriting systems (this is not Büchi's terminology) generate just the regular term languages. More details can be found in a paper I wrote on the book, mentioned in the appendix.

2. Discrete Spaces

A finite automaton is a discrete deterministic system. Starting with his doctoral dissertation and continuing through all his life, Richard Büchi also used the more general setting of discrete spaces to investigate discrete systems. Some of this work on set spaces, lattices, discrete linear and convex closure, directed graphs, and matroids is contained in the publications nos. 1–4, 19, and 34–38 William Fenton gave a talk on his work with Richard Büchi on convexity theory at this 1985 conference. Other material will likely remain unpublished.

3. Finite Automata on Infinite Input, Monadic Theories, and Determinacy

The (weak) monadic second-order theory of a class of structures is the elementary theory enriched by quantifiable variables for (finite) predicates, i.e., sets. It was Richard Büchi's idea to express statements on finite automata as formulas of (weak) monadic second-order theories, and in this way to get decidability and definability results for such theories by using combinatorial statements expressed in terms of finite automata. For an example write T (true) for the state "odd" and the input 1 and F (false) for the state "even" and the input 2 of the automaton in Section 1:

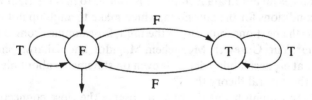

Then the automaton accepts the word $w \in \{F, T\}^*$ iff there is an accepting "run," i.e., a sequence of states $v \in \{F, T\}^*$ beginning and ending with F and following the transition condition. This statement can be read as a formula of the weak monadic second order theory of the successor on the natural numbers:

$$\exists Z. \neg Z_0 \wedge \forall t(Zt' \leftrightarrow (Zt \leftrightarrow Xt)),$$

where $X := wFF \ldots$, $Z := vFF \ldots$, and $'$ is the successor function. Richard Büchi showed that "automata formulas" of this type constitute a normal form and in this way proved that the theory is decidable; see publication no. 7. The same method works for arbitrary ordinals (publication no. 13). Using this method Calvin Elgot obtained related results; see also the joint abstract no. 5.

The input X and the run Z in the above automata formula are actually infinite, though eventually constant, sequences. Two years later Richard Büchi had generalized his method to automata on ω-input (ω-automata) and to the strong monadic second order successor theory (publications nos. 8 and 16). For example, the formula

$$\exists Z . \neg Z_0 \wedge \forall t(Zt' \leftrightarrow (Zt \leftrightarrow Xt)) \wedge \exists^\omega t\, Zt$$

says that the above automaton accepts those sequences X which contain an even number of Fs or infinitely many of them. Automata formulas for non-deterministic ω-automata yield a normal form:

$$\exists Z . A[Z_0] \wedge \forall t B[Xt, Zt, Zt'] \wedge \exists^\omega t\, C[Zt].$$

Robert Mcnaughton proved that every such automaton can be changed into a deterministic one if one replaces the acceptance condition

$$\exists^\omega t\, C[Zt] \qquad \text{by} \qquad \sup Z \in D \qquad \text{where} \quad \sup Z := \{s;\, \exists^\omega t\, Zt = s\}$$

is the set of states which occur infinitely often.

Similar ideas with more and more sophisticated automata and thus more and more help from infinite combinatorics and set theory work for all ordinals $<\omega_2$; see publications nos. 14, 24, 28. They also work for nonstandard structures like $\omega + \omega^*$, and allow one to axiomatize the theories in question and to characterize their complete extensions by prime models (publications nos. 25 and 26); see also the paper by Charles Zaiontz in the same volume as no. 28. The central part of the normal form construction is always a complementation lemma, which says that the automaton-acceptable sets are closed under complement, and thus allows one to rewrite the negation of an automata formula as an automata formula. It is crucial to find the right transition and final conditions for the automata, which make them jump not only from one state to the next one, but also to the limit, for example, from ω to $\omega + 1$, or to $\omega + \omega^*$. Yuri Gurevich, Menachem Magidor, and Saharon Shelah have shown that at ω_2 decidability depends even on statements about big ordinals; so what is the natural theory there?

Robert McNaughton was the first to observe the close connection of the area to problems on infinite two-person games with complete information. Such a game is given by a condition $C(X, Y)$ on ω-sequences over two finite sets H and K. Two players I and J alternately choose from H and K respectively. Their moves depend on the previous choices of both players, and constitute two ω-sequences X and Z. Player J wins if the result satisfies the condition C; otherwise player I wins. A winning strategy for player J (I) is an operation W (V) which to any play of the opponent yields a play which beats him:

$$\forall X \, C(X, WX), \quad \text{or} \quad \forall Y \, \neg C(VY, Y) \quad \text{respectively.}$$

A game is determinate if one of the players has a winning strategy. A finite-state winning strategy is one which can be realized by an ω-automaton.

Richard Büchi and Lawrence Landweber proved that all games definable in the monadic second-order theory of successor are determinate; one can decide who wins, and can construct a finite-state winning strategy; see publications nos. 15 and 18. This had been stated as problem by Richard Büchi in no. 8, and conjectured by Robert McNaughton. By another result of Büchi and Landweber (see no. 16), the games in question are just the regular sets in the Boolean algebra over F_σ, i.e., definable by Boolean combinations of formulas

$$\exists y \forall t \geq y \, (\bar{X}t, \bar{Y}t) \in L,$$

where (X, Y) is thought of as a path in an $H \times K$-tree, $(\bar{X}t, \bar{Y}t)$ is the corresponding node at height t, and the congruence induced by L has finite rank (cf. Section 1). In the early 1970s Richard Büchi removed the restriction to regular sets; see nos. 20 (for closed sets) and 27. The last part of the result then reads: one can decide who wins, and can construct a winning strategy which is finite-state over the given game. He published the proof in no. 29 for the slightly bigger class of $F_{\sigma\delta} \cap G_{\delta\sigma}$-games (one more quantifier change).

In a very difficult and complicated construction Michael Rabin extended Richard Büchi's method of handling monadic second-order theories with the help of finite automata, to the binary tree, i.e., to the case of two successors. He observed that his proof together with McNaughton's result on deterministic ω-automata yields the second half of the Büchi–Landweber result, namely decidability and construction; it does not yield determinacy. In no. 27 Richard Büchi introduced determinacy as a general tool for eliminating quantifiers, and proved especially that his determinacy for Boolean F_σ-games together with McNaughton's result yields Rabin's decision procedure. For this he showed that in the monadic second-order theory of two successors the complementation lemma follows from a determinacy statement, namely, the automata normal form here is

$$\exists W . A[We] \wedge \forall u \, B[Vu, Wu, Wu1, Wu2] \wedge \forall X \sup WX \in D, \tag{1}$$

where the input V and the run W are finite-state trees, X is a path through the tree, and sup WX is the sup (as defined above) of the run W along the path X. Using McNaughton's result he brought formula (1) into the form

$$\exists W \forall X \exists Z . Z_0 = c \wedge \forall t (Zt' \leftrightarrow B[Zt, Xt, W\bar{X}t, V\bar{X}t]) \wedge \sup Z \in \tilde{D}, \tag{2}$$

where V, W, and X are as above, and Z is the run of a deterministic ω-automaton along X. Now the formula

$$\exists Z . Z_0 = c \wedge \forall t (Zt' \leftrightarrow B[Zt, Xt, Yt, V\bar{X}t]) \wedge \sup Z \in \tilde{D}$$

defines a special Boolean F_σ-game where player I chooses from the set $\{1, 2\}$, and V is a parameter tree. Therefore formula (2) says that player J has a winning strategy for that game. By determinacy the negation of (2) can be brought into the same form; thus, this holds for (1) as well.

4. Quadratic Forms, the Five Squares Problem, and Diophantine Equations

For many years Richard Büchi worked on quadratic forms. He presented Gauss' class field theory in an algorithmic form, rather in the spirit of Gauss himself, but more thoroughly organized in a way which was suitable for his purpose. It seems that his interest in quadratic forms stemmed from Hilbert's 10th problem, which asks for a procedure to decide whether a given set of Diophantine (i.e., integer coefficient) equations has a solution. He tried to approach it by reducing Diophantine equations to equations on square numbers in the following way.

Let $x_1 < \cdots < x_n$ be positive integers, $n \geq 3$. If they are consecutive, i.e., $x_{i+1} = x_i + 1$, then the second difference of their squares in 2, namely,

$$(x_{i+1}^2 - x_i^2) - (x_i^2 - x_{i-1}^2) = (x_i + 1)^2 - 2x_i^2 + (x_i - 1)^2 = 2.$$

Richard Büchi raised the question: Are there nonconsecutive integers with this property? He knew a procedure to generate infinitely many such sequences in the case $n = 4$ (and thus $n = 3$); an example is 6, 23, 32, 39. For $n = 5$ the question is open; his students called it the "5-squares problem." He proved: If for some n the answer to the n-squares problem is "no," then for any Diophantine equation P one can construct a set Q of Diophantine equations with only quadratic terms s.t. P has an integral solution iff Q has:

$$\exists z_1 \ldots z_m P(z_1, \ldots, z_m) = 0 \quad \text{iff} \quad \exists y_1 \ldots y_k \; A \begin{pmatrix} y_1^2 \\ \vdots \\ y_k^2 \end{pmatrix} = B,$$

where P is a polynomial with integer coefficients, A is an integer matrix, and B is an integer vector. The idea of the proof is to define the relations

$$y = x + 1, \quad y = x^2, \quad \text{and finally} \quad z = x \cdot y,$$

with the help of squared variables and existential quantifiers. Thus if the answer is "no," Hilbert's 10th problem can be stated in this much simpler form.

Since by now Hilbert's 10th problem has been solved in the negative, the result amounts to the following: If the answer to some n-squares problem is "no," then there is no procedure to decide whether a given set of Diophantine equations with only squared variables has a solution; or slightly more general: then the existential theory of addition over the integers with the predicate "is a square" is undecidable.

Carl Friedrich Gauss in his *Disquisitiones Arithmeticae* (1800) investi-

which numbers M can be represented by quadratic forms, i.e., as

$$M = ax^2 + bxy + cy^2,$$

where a, b, c and x, y are integers. He denotes a form F by its coefficients (a, b, c), and classifies forms by the determinant of their matrix

$$D := ac - \frac{b^2}{4} = \begin{vmatrix} a & b/2 \\ b/2 & c \end{vmatrix},$$

or, as one does it today, by their discriminant $b^2 - 4ac = -4D$. He calls two forms F and F' equivalent if they can be transformed into each other by a unimodular transformation:

$$F' = A^{\mathsf{T}}FA = \begin{pmatrix} h & k \\ i & j \end{pmatrix}\begin{pmatrix} a & b/2 \\ b/2 & c \end{pmatrix}\begin{pmatrix} h & i \\ k & j \end{pmatrix} \quad \text{where} \quad |A| = hj - ki = \pm 1.$$

Equivalent forms have the same discriminant and represent the same numbers, and conversely: two forms which represent the same numbers are equivalent. For each discriminant there are only finitely many classes of forms (class field theorem). Gauss devotes much of the book to studying which types of classes occur for which types of discriminants.

For these investigations Richard Büchi introduces three structures. First he notes that the positive unimodular matrices (of dimension 2 by 2) are freely generated from the unit matrix e through multiplication from the right by matrices 1 and 2, where

$$e := \begin{pmatrix} 1 & 0 \\ 0 & 1 \end{pmatrix}, \quad 1 := \begin{pmatrix} 1 & 0 \\ 1 & 1 \end{pmatrix}, \quad 2 := \begin{pmatrix} 1 & 1 \\ 0 & 1 \end{pmatrix}.$$

Thus they can be represented by the full binary tree of Section 1. To generate all unimodular matrices he adds the inverses

$$\dot{1} = \begin{pmatrix} 1 & 0 \\ -1 & 1 \end{pmatrix} \quad \text{and} \quad \dot{2} = \begin{pmatrix} 1 & -1 \\ 0 & 1 \end{pmatrix}$$

of 1 and 2. In this way he gets the structure $\mathfrak{U}$

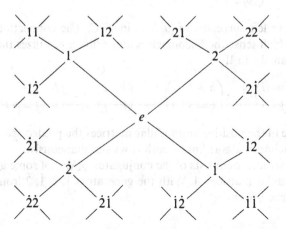

of unimodular matrices. It is not free; for example, $L^2 = -e$, and thus $L^4 = e$, where

$$L := 2\dot{1}2 = \begin{pmatrix} 0 & 1 \\ -1 & 0 \end{pmatrix}.$$

Gauss uses L instead of 1 to generate the unimodular matrices. Therefore he misses the connection to the word algebra, and thus to the other two structures.

Richard Büchi gets the second structure from an algorithm which Nicomachus introduced around 100 A.D. in his *Arithmetics*. Nicomachus states that it generalizes the Euclidean algorithm for computing the greatest common divisor of a pair of integers. It applies to triples of integers, and yields at each step four new triples. Thus it results in the structure NIC:

$$\begin{array}{ccc} \diagdown & & \diagup \\ (a + b + c, b + 2c, c) & & (a, b + 2a, a + b + c) \\ \diagup & \diagdown \quad (a, b, c) \quad \diagup & \diagdown \\ (a, b - 2a, a - b + c) & & (a - b + c, b - 2c, c) \\ \diagup & \diagdown & \diagdown \end{array}$$

If $F(a, b, c)$ is a form, then the NIC-structure consists of the forms equivalent to F. In fact NIC is a homomorphic image of the structure $\mathfrak{U}$ of unimodular matrices if we map the matrix A into the form $A^{\mathrm{T}} F A$ (see above). For example, $e^{\mathrm{T}} F e = F$ and

$$1^{\mathrm{T}}(a, b, c)1 = \begin{pmatrix} 1 & 1 \\ 0 & 1 \end{pmatrix}\begin{pmatrix} a & b/2 \\ b/2 & c \end{pmatrix}\begin{pmatrix} 1 & 0 \\ 1 & 1 \end{pmatrix} = \begin{pmatrix} a + b/2 & b/2 + c \\ b/2 & c \end{pmatrix}\begin{pmatrix} 1 & 0 \\ 1 & 1 \end{pmatrix}$$

$$= \begin{pmatrix} a + b + c & b/2 + c \\ b/2 + c & c \end{pmatrix} = (a + b + c, b + 2c, c),$$

which is upper left successor of (a, b, c) in NIC. The connection to $\mathfrak{U}$ also explains in which sense the Nicomachus algorithm generalizes the Euclidean algorithm, namely, in $\mathfrak{U}$

$$\begin{pmatrix} h & i \\ k & j \end{pmatrix} \cdot \dot{1} = \begin{pmatrix} h - i & i \\ k - j & j \end{pmatrix}, \qquad \begin{pmatrix} h & i \\ k & j \end{pmatrix} \cdot \dot{2} = \begin{pmatrix} h & i - h \\ k & j - k \end{pmatrix},$$

and therefore in the positive unimodular matrices the predecessor operations apply the Euclidean algorithm to both rows simultaneously.

The third structure consists of the conjugates $\dot{A}X A$ of some given matrix X for unimodular matrices A. With the generators $1, 2, \dot{1}, \dot{2}$ from above the conjugacy structure is

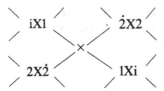

It is obviously again a homomorphic image of the structure $\mathfrak{U}$. In fact, it is isomorphic to the NIC-structure if we map the form $F = (a, b, c)$ into the matrix

$$X := \begin{pmatrix} \frac{1}{2}(\iota - b) & -c \\ a & \frac{1}{2}(\iota + b) \end{pmatrix},$$

where $\iota = 0, 1$ is the residuum modulo 4 of the discriminant of F. Thus also the conjugacy structure can be viewed as representing the class of forms equivalent to F.

Going back and forth between the three structures and using whichever he found convenient, Richard Büchi reconstructed Gauss' results on quadratic forms, and worked on the 5-squares problem. Nothing of this work is published, and is preserved only in very many notes, some of it in lecture notes taken by the author and by Steven Senger.

5. Words, Turing Machines, and Recursive Functions

Richard Büchi refers often to Hilbert's 10th problem in his early notes. It may well be that more of his work was stimulated by that problem, and this is surely the case with the existential theory of words over a finite alphabet. Since we represent numbers by words, Hilbert's 10th problem would be solved in the negative if the existential theory of words were undecidable. He therefore was long interested in that problem, but was overtaken by Makanin who proved in 1976 that this theory is decidable. An offspin of this interest is the work he did with Steven Senger on definability and decidability in extensions of that theory; see publications nos. 32 and 33. Steven Senger gave a talk on these results at this 1985 conference.

Besides the first accounts of his work on finite automata and algebra (no. 10) and on monadic second-order theories (no. 8) Richard Büchi published a third landmark paper in 1962 (no. 9), in which he proves that the prefix $\exists \wedge \forall\exists\forall$ is a reduction type. In the proof he expresses the halting problem for Turing machines in formulas of this type, and thereby—although he did not invent "domino" or "tiling" problems—he is the first to use them in such a reduction, which is now the standard method.

Also in the area of computability, in 1980 Richard Büchi (together with Bernd Mahr) joined the present author in studying recursive definitions on arbitrary partial algebras. In publications nos. 30 and 31 we show that recursion is a natural tool to formulate algorithms over any data structure, and to analyze their cost.

6. Abstraction

The relation of proportionality in a multiplicative group,

$$\rho(x, y, u, v) :\leftrightarrow x \cdot y^{-1} = u \cdot v^{-1},$$

can be characterized by simple axioms. Let T be the elementary theory of ρ with these axioms, let T_1 be the elementary theory of groups with multiplication $\cdot$, inverse^{-1}, and unit e as primitives, and let T_2 be T plus the constant e without additional axioms. Then T_1 and T_2 are strongly equivalent; namely their basic notions are interdefinable, and thus they have the same models up to this change of primitives. Richard Büchi and Jesse B. Wright in publications nos. 5 and 6 call T an abstraction of T_2 (and thus of group theory) in either of the following senses:

(i) T is weaker than T_2 in expressive power, namely the constant e is not definable in T; but T and T_2 have the same models up to adding or deleting e.

(ii) The automorphism group of T extends that of T_2, namely by the translations (which are not group automorphisms).

Another example of abstraction appears already in Richard Büchi's doctoral thesis, no. 1. These observations were the starting point for his long-lasting work towards a theory of definability, incorporating Craig's lemma and Beth's definability theorem. Part of the work was done together with Kenneth Danhof, and is published in nos. 21–23. The manuscript no. 40 was not published because Büchi refused to shorten it; see abstract no 11.

Acknowledgment

I am grateful to Sylvia Büchi, Leonard Lipshitz, and Walter Schnyder who helped me through the ups and downs of discovering what Richard Büchi had left behind. I also thank William Fenton and Steven Senger, his two last doctoral students, for helpful conversations, and Ernst Specker for arranging with the ETH Zürich to store the Büchi material. Beat Glaus of the Archiv of the ETH library has gone through the tremendous work of cataloging it, so that now all the notes and unpublished papers are accessible.

Publications

1. Die Boole'sche Partialordnung und die Paarung von Gefügen. *Portugal. Math.* **7** (1950), 119–178. Doctoral dissertation.

2. Representation of complete lattices by sets. *Portugal. Math.* **11** (1952), 151–167.

3. Investigation of the equivalence of the axiom of choice and Zorn's lemma from the viewpoint of the hierarchy of types. *J. Symbolic Logic* **18** (1953), 125–135.

4. On the existence of totally heterogeneous spaces. *Fund. Math.* **41** (1954), 97–102.

5. (with Jesse B. Wright) The theory of proportionality as an abstraction of group theory. *Math. Ann.* **130** (1955), 102–108.

6. (with Jesse B. Wright) Invariants of the anti-automorphisms of a group. *Proc. Amer. Math. Soc.* **8** (1957), 1134–1140.

7. Weak second-order arithmetic and finite automata. *Z. Math. Logik Grundlag. Math.* **6** (1960), 66–92.

8. On a decision method in restricted second order arithmetic. Proc. Int. Congress Logic, Methodology, and Philosophy of Science, 1960, Berkeley. Stanford University Press, Stanford, California, 1962, 1–11. Invited address.

9. Turing machines and the Entscheidungsproblem. *Math. Ann.* **148** (1962), 201–213.

10. Mathematische Theorie des Verhaltens endlicher Automaten. *Z. Angew. Math. Mech.* **42** (1962), T9–T16. Invited address, yearly meeting of the German Soc. Appl. Math. and Mech.

11. Regular canonical systems and finite automata. *Arch. Math. Logik Grundlag.* **6** (1964), 91–111.

12. Algebraic theory of feedback in discrete systems. In: E. Caiamello (ed.), *Automata Theory*, 1. Course on Automata Theory, Ravello, Italy, 1964, Academic Press, New York, 1966, 70–101.

13. Transfinite automata recursions and weak second order theory of ordinals. Proc. Int. Congress Logic, Methodology, Philosophy of Science 1964, Jerusalem. North-Holland, Amsterdam, 1965, 2–23. Invited address.

14. Decision methods in the theory of ordinals. *Bull. Amer. Math. Soc.* **71** (1965), 767–770.

15. (with Lawrence H. Landweber) Solving sequential conditions by finite state strategies. *Trans. Amer. Math. Soc.* **138** (1969), 295–311.

16. (with Lawrence H. Landweber) Definability in the monadic second-order theory of successor. *J. Symbolic Logic* **34** (1969), 166–170.

17. (with William H. Hosken) Canonical systems which produce periodic sets. *Math. Systems Theory* **4** (1970), 81–90.

18. Algorithmisches Konstruieren von Automaten und die Herstellung von Gewinnstrategien nach Cantor-Bendixson. Tagungsbericht Automatentheorie und Formale Sprachen, Math. Forschungsinst. Oberwolfach, 1969. Mannheim, Germany, 1970, pp. 385–398. Invited address.

19. (with Gary Haggard) Jordan circuits of a graph. *J. Combin. Theory* **10** (1971), 185–197.

20. (with Steven Klein) On the presentation of winning strategies via the Cantor-Bendixson Method. Report Purdue University CSD TR-81, 1972, 14 pp.

21. (with Kenneth J. Danhof) Model theoretic approaches to definability. *Z. Math. Logik Grundlag. Math.* **18** (1972), 61–70.

22. (with Kenneth J. Danhof) Definability in normal theories. *Israel J. Math.* **14** (1973), 248–256.

23. (with Kenneth J. Danhof) Variations on a theme of Cantor in the theory of relational structures. *Z. Math. Logik Grundlag. Math.* **19** (1973), 411–426.

24. The monadic second-order theory of ω_1. In: J.R. Büchi, D. Siefkes, *The Monadic Second-Order Theory of All Countable Ordinals.* Lecture Notes in Math, vol. 328, Springer-Verlag, Berlin, 1975, pp. 1–127.

25. (with Dirk Siefkes) Axiomatization of the monadic second-order theory of ω_1. Same volume as 24, pp. 129–217.
26. (with Dirk Siefkes) The complete extensions of the monadic second-order theory of countable ordinals. Report Forschungsinstitut für Math., ETH Zürich (1974), 45 pp. Z. Math. Logik Grundlag. Math. 29 (1983), 289–312.
27. Using Determinacy of Games to Eliminate Quantifiers. In: M. Karpinski (ed.) Fund. of Comp. Theory, 1977. Lecture Notes in Comp. Sci., vol. 56, Springer-Verlag, Berlin, pp. 367–378. Invited address.
28. (with Charles Zaiontz) Deterministic automata and the monadic theory of ordinals $<\omega_2$. Z. Math. Logik Grundlag. Math. 29 (1983), 313–336.
29. State-strategies for games in $F_{\sigma\delta} \cap G_{\delta\sigma}$. J. Symbolic Logic 48 (1983), 1171–1198.
30. (with Bernd Mahr, Dirk Siefkes) Manual on REC—A Language for Use and Cost Analysis of Recursion Over Arbitrary Data Structures. Bericht Nr. 84-06 Techn. Univ. Berlin, FB Informatik (1984), 79 pp.
31. (with Bernd Mahr, Dirk Siefkes) Recursive Definition and Complexity of Functions Over Arbitrary Data Structures. Proc. 2. Frege Conference, G. Wechsung ed., Akademie Verlag, Berlin, 1984, pp. 303–308.
32. (with Steven O. Senger) Coding in the existential theory of concatenation. Arch. Math. Logik Grundlag. 26 (1986/87), 101–106.
33. (with Steven O. Senger) Existential definability and decidability of extensions of the theory of concatenation. Z. Math Logik Grundlag. Math. 34 (1988), 337–342.
34. (with William E. Fenton) Large convex sets in oriented matroids. J. Combin. Theory, series B 45 (1988), 293–304.
35. (with William E. Fenton) Directed circuits of a graph with an application to series-parallel graphs. Manuscript.
36. Implicative Boolean algebras, Copeland's theory of conditional probability. Manuscript.
37. (with T. Michael Owens) Skolem rings and their varieties. Report Purdue University CSD TR-140, 1975.
38. (with T. Michael Owens) Complemented monoids and hoops. To be submitted.
39. Finite automata, their algebras and grammars. An approach to a theory of formal expressions. D. Siefkes (ed.), Springer-Verlag, New York, 1989.
40. Relatively categorical and normal theories. Manuscript, 1963–1965.

Abstracts

1. (with Jesse B. Wright) Abstraction versus generalization. Proc. Int. Congress of Math. Amsterdam 1954 , 398.
2. (with Jesse B. Wright) A fundamental concept in the theory of models. J. Symbolic Logic 21 (1956), 110.
3. (with William Craig) Notes on the family PC_Δ of sets of models. J. Symbolic Logic 21 (1956), 222–223.
4. (with Calvin C. Elgot, and Jesse B. Wright) The nonexistence of certain algorithms of finite automata theory. Notices Amer. Math. Soc. 5, 1 (1958), 98.
5. (with Calvin C. Elgot) Decision problems of weak second-order arithmetic and finite automata, I. Notices Amer. Math. Soc. 5, 7 (1958), 834.
6. Regular canonical systems and finite automata. Notices Amer. Math. Soc. 6, 6 (1959), 618.
7. On the hierarchy of monadic predicate quantifiers. Notices Amer. Math. Soc. 7

(1960), 381.

8. On a problem of Tarski. *Notices Amer. Math. Soc.* **7** (1960), 382.
9. Turing machines and the Entscheidungsproblem. *Notices Amer. Math. Soc.* **8** (1961), 354.
10. Validity in finite domains. *Notices Amer. Math. Soc.* **8** (1961), 354.
11. Logic, arithmetic, and automata. Proc. Int. Congress of Math. 1963. Almquist and Wiksells, Uppsala 1963.
12. Relatively categorical and normal theories. Proc. Int. Symp. on Theory of Models, Berkeley, 1963. North-Holland, Amsterdam, 1965, pp. 424–426.
13. Transfinite automata recursions. *Notices Amer. Math. Soc.* **12** (1965), 371.
14. Restricted second-order theory of ordinals. *Notices Amer. Math. Soc.* **12** (1965), 457.
15. (with Lawrence H. Landweber) Definability in the monadic second-order theory of successor. *Notices Amer. Math. Soc.* **14** (1967), 852.
16. Affine definability of affine invariants of Euclidean geometry. *Notices Amer. Math. Soc.* **15** (1968), 932.
17. (with Kenneth J. Danhof) A strong form of Beth's definability theorem. *Notices Amer. Math. Soc.* **15** (1968), 932.
18. (with Dirk Siefkes) Axiomatization of the monadic second-order theory of countable ordinals. Tarski Symposium, Berkeley, 1971.
19. The monadic second-order theory of ω_1. Preliminary report. *Notices Amer. Math. Soc.* **18** (1971), 662.

Doctoral Students

William Henry Hosken (1966)—Canonical Systems which Produce Regular Sets.
Lawrence Hugh Landweber (1967)—A Design Algorithm for Sequential Machines and Definability in Monadic Second Order Arithmetic.
Gary Martin Haggard (1968)—Embedding of Graphs in Surfaces.
Kenneth Joe Danhof (1969)—On Definability and the Cantor Method in Model Theory.
Peng-Siu Mei (1971)—Linear Closure Spaces and Matroids, Convex Closure Spaces and Paramatroids.
Jean-Louis Lassez (1973)—On the Relationship Between Prefix Codes, Trees and Automata.
Charles Zaiontz (1974)—Automata and the Monadic Theory of Ordinals $< \omega_2$.
Terrence Michael Owens (1975)—Varieties of Skolem Rings.
William Ellis Fenton (1982)—Axiomatic Convexity Theory.
Steven Orville Senger (1982)—The Existential Theory of Concatenation over a Finite Alphabet.

Further Publications

Dirk Siefkes: Grammars for Terms and Automata. On a book by the late J. Richard Büchi. In: *Computation Theory and Logic*, Egon Börger (ed.), Lect. Notes Comp. Sci., vol. 270 (Rödding memorial volume), Springer-Verlag, New York, 1987, pp. 349–359.
Saunders Maclane, Dirk Siefkes ed.: *Collected Works of J. Richard Büchi*. Springer-Verlag, New York, 1990.

Index Nominum